Active Organic and Organic-Inorganic Hybrid Coatings and Thin Films

Active Organic and Organic-Inorganic Hybrid Coatings and Thin Films

Challenges, Developments, Perspectives

Editors

Assunta Marrocchi
Maria Laura Santarelli

MDPI • Basel • Beijing • Wuhan • Barcelona • Belgrade • Manchester • Tokyo • Cluj • Tianjin

Editors
Assunta Marrocchi
University of Perugia
Italy

Maria Laura Santarelli
University of Rome Sapienza
Italy

Editorial Office
MDPI
St. Alban-Anlage 66
4052 Basel, Switzerland

This is a reprint of articles from the Special Issue published online in the open access journal *Coatings* (ISSN 2079-6412) (available at: https://www.mdpi.com/journal/coatings/special_issues/org_hybrid).

For citation purposes, cite each article independently as indicated on the article page online and as indicated below:

LastName, A.A.; LastName, B.B.; LastName, C.C. Article Title. *Journal Name* **Year**, *Article Number*, Page Range.

ISBN 978-3-03936-852-5 (Hbk)
ISBN 978-3-03936-853-2 (PDF)

Cover image courtesy of Assunta Marrocchi.

© 2020 by the authors. Articles in this book are Open Access and distributed under the Creative Commons Attribution (CC BY) license, which allows users to download, copy and build upon published articles, as long as the author and publisher are properly credited, which ensures maximum dissemination and a wider impact of our publications.
The book as a whole is distributed by MDPI under the terms and conditions of the Creative Commons license CC BY-NC-ND.

Contents

About the Editors vii

Preface to "Active Organic and Organic-Inorganic Hybrid Coatings and Thin Films" ix

Mohammad Mizanur Rahman, Md. Hasan Zahir, Md. Bashirul Haq, Dhafer A. Al Shehri and A. Madhan Kumar
Corrosion Inhibition Properties of Waterborne Polyurethane/Cerium Nitrate Coatings on Mild Steel
Reprinted from: *Coatings* **2018**, *8*, 34, doi:10.3390/coatings8010034 1

Marco F. D'Elia, Andreas Braendle, Thomas B. Schweizer, Marco A. Ortenzi, Stefano P. M. Trasatti, Markus Niederberger and Walter Caseri
Poly(Phenylene Methylene): A Multifunctional Material for Thermally Stable, Hydrophobic, Fluorescent, Corrosion-Protective Coatings
Reprinted from: *Coatings* **2018**, *8*, 274, doi:10.3390/coatings8080274 13

Chijia Wang, Huaiyuan Wang, Yue Hu, Zhanjian Liu, Chongjiang Lv, Yanji Zhu and Ningzhong Bao
Anti-Corrosive and Scale Inhibiting Polymer-Based Functional Coating with Internal and External Regulation of TiO_2 Whiskers
Reprinted from: *Coatings* **2018**, *8*, 29, doi:10.3390/coatings8010029 25

Simone Giaveri, Paolo Gronchi and Alessandro Barzoni
IPN Polysiloxane-Epoxy Resin for High Temperature Coatings: Structure Effects on Layer Performance after 450 °C Treatment
Reprinted from: *Coatings* **2017**, *7*, 213, doi:10.3390/coatings7120213 41

Caterina Lesaint Rusu, Malin Brodin, Tor Inge Hausvik, Leif Kåre Hindersland, Gary Chinga-Carrasco, Mari-Ann Einarsrud and Hilde Lea Lein
The Potential of Functionalized Ceramic Particles in Coatings for Improved Scratch Resistance
Reprinted from: *Coatings* **2018**, *8*, 224, doi:10.3390/coatings8060224 57

Dongkyu Kim and Choongik Kim
A Ladder-Type Organosilicate Copolymer Gate Dielectric Materials for Organic Thin-Film Transistors
Reprinted from: *Coatings* **2018**, *8*, 236, doi:10.3390/coatings8070236 69

Sadaf Bashir Khan, Hui Wu and Zhengjun Zhang
Omnidirectional SiO_2 AR Coatings
Reprinted from: *Coatings* **2018**, *8*, 210, doi:10.3390/coatings8060210 79

Riccardo Narducci, Maria Luisa Di Vona, Assunta Marrocchi and Giorgio Baldinelli
Stabilized SPEEK Membranes with a High Degree of Sulfonation for Enthalpy Heat Exchangers
Reprinted from: *Coatings* **2018**, *8*, 190, doi:10.3390/coatings8050190 93

Francesca Sbardella, Lucilla Pronti, Maria Laura Santarelli, José Marìa Asua Gonzàlez and Maria Paola Bracciale
Waterborne Acrylate-Based Hybrid Coatings with Enhanced Resistance Properties on Stone Surfaces
Reprinted from: *Coatings* **2018**, *8*, 283, doi:10.3390/coatings8080283 109

Mariaenrica Frigione and Mariateresa Lettieri
Novel Attribute of Organic–Inorganic Hybrid Coatings for Protection and Preservation of Materials (Stone and Wood) Belonging to Cultural Heritage
Reprinted from: *Coatings* **2018**, *8*, 319, doi:10.3390/coatings8090319 . . . . **123**

About the Editors

Assunta Marrocchi is a professor of Chemistry at the University of Perugia (Italy), with expertise in organic synthesis. Her research interests include renewable resources, materials design and sustainable synthesis, heterogeneous catalysis, polymer chemistry, and safe chemicals. Marrocchi obtained her Ph.D. in Chemical Science from the University of Perugia. She has published over 100 peer-reviewed scientific papers and two patents. She can be reached by e-mail at assunta.marrocchi@unipg.it

Maria Laura Santarelli is an assistant professor at the University of Rome "Sapienza", with research interests in the area of materials characterization, polymer science, conservation of cultural heritage, petroleum derivatives, photocatalytic materials, and applications of graphene. Santarelli earned her Ph.D. in Materials, Raw Materials, and Metallurgy at the University of Rome "Sapienza". She has published over 70 papers in national and international journals. She can be reached by e-mail at: marialaura.santarelli@uniroma1.it

Preface to "Active Organic and Organic-Inorganic Hybrid Coatings and Thin Films"

During the past years, "smart" coatings and thin films have been widely explored in terms of structure material properties, processing, morphology, and performance. This Special Issue of Coatings is dedicated to this exciting research field, focusing on the latest developments in the design and synthesis of smart organic and organic–inorganic hybrid coatings and thin films. To reflect the broad scope of this topic, contributions from leading scientists across the world, whose research addresses smart coatings and thin films but from different perspectives and in varying contexts, have been gathered. Rahman and Baq (A. Corrosion Inhibition Properties of Waterborne Polyurethane/Cerium Nitrate Coatings on Mild Steel, https://www.mdpi.com/2079-6412/8/1/34) focused on the preparation of stable waterborne polyurethane/cerium nitrate hybrid dispersions, which proved to be efficient for corrosion protection of mild steel. Caseri described the preparation and properties of new poly(phenylene methylene)-based coatings with anti-corrosion activity for metal alloys exposed to NaCl solutions (Poly(Phenylene Methylene): A Multifunctional Material for Thermally Stable, Hydrophobic, Fluorescent, Corrosion-Protective Coatings, https://www.mdpi.com/2079-6412/8/8/274). Wang (Anti-Corrosive and Scale Inhibiting Polymer-Based Functional Coating with Internal and External Regulation of TiO2 Whiskers, https://www.mdpi.com/2079-6412/8/1/29) focused on functional epoxy-TiO2 and silicone-epoxide-TiO2 composite coatings with corrosion and scale-inhibiting properties for steel substrates. Gronchi (IPN Polysiloxane-Epoxy Resin for High Temperature Coatings: Structure Effects on Layer Performance after 450 °C Treatment, https://www.mdpi.com/2079-6412/7/12/213) presents the synthesis and characterization of silicone-epoxy-acrylic interpenetrating networks (IPNs) and IPN-graphene nanocomposite coatings against the thermo-oxidative corrosion of, e.g., the metal parts of the internal combustion engines, turbines, and heaters. Lein (The Potential of Functionalized Ceramic Particles in Coatings for Improved Scratch Resistance, https://www.mdpi.com/2079-6412/8/6/224) focused on melamine formaldehyde–alumina hybrids to enhance the mechanical properties, particularly the scratch resistance, of floor laminates. Kim (A Ladder-Type Organosilicate Copolymer Gate Dielectric Materials for Organic Thin-Film Transistors, https://www.mdpi.com/2079-6412/8/7/236) focused on dielectric thin films based on a new class of ladder-type organosilicate polymers, which proved to be curable at low temperatures. Zhang describes a new approach for the design and fabrication of SiO2-based antireflective coating for optoelectronic applications. Narducci (Stabilized SPEEK Membranes with a High Degree of Sulfonation for Enthalpy Heat Exchangers, https://doi.org/10.3390/coatings8050190) focused on cross-linked sulfonated poly(ether ether ketone) (SPEEK) membranes for application in enthalpy heat exchanger systems, enabling high level of sensible heat exchange, and remarkable variation in the water vapor transfer between individual air flows. Sbardella and Bracciale (Waterborne Acrylate-Based Hybrid Coatings with Enhanced Resistance Properties on Stone Surfaces, https://www.mdpi.com/2079-6412/8/8/283) described the green synthesis and properties of waterborne nanostructured hybrid silica/polyacrylate coatings for application in the preservation of built heritage. The review by Frigione (Novel Attribute of Organic–Inorganic Hybrid Coatings for Protection and Preservation of Materials (Stone and Wood) Belonging to Cultural Heritage, https://www.mdpi.com/2079-6412/8/9/319) examines recent advances in hybrid smart nanocomposite coatings for the conservation treatment of stone-

and wood-built cultural heritage. We are confident that the articles contained in this Special Issue will serve to further stimulate advances in this research area. We thank all our friends and colleagues who contributed papers to the themed Issue.

Assunta Marrocchi, Maria Laura Santarelli
Editors

Article

Corrosion Inhibition Properties of Waterborne Polyurethane/Cerium Nitrate Coatings on Mild Steel

Mohammad Mizanur Rahman [1,*], Md. Hasan Zahir [2], Md. Bashirul Haq [3,*], Dhafer A. Al Shehri [3] and A. Madhan Kumar [1]

1. Center of Research Excellence in Corrosion, King Fahd University of Petroleum and Minerals, Dhahran 31261, Saudi Arabia; madhankumar@kfupm.edu.sa
2. Center of Research Excellence in Renewable Energy, King Fahd University of Petroleum and Minerals, Dhahran 31261, Saudi Arabia; hzahir@kfupm.edu.sa
3. Department of Petroleum Engineering, College of Petroleum and Geosciences, King Fahd University of Petroleum and Minerals, Dhahran 31261, Saudi Arabia; alshehrida@kfupm.edu.sa
* Correspondence: mrahman@kfupm.edu.sa (M.M.R.); bhaq@kfupm.edu.sa (M.B.H.); Tel.: +966-13-860-7210 (M.M.R.)

Received: 26 September 2017; Accepted: 10 January 2018; Published: 15 January 2018

Abstract: Waterborne polyurethane (WBPU)/cerium nitrate (Ce(NO$_3$)$_3$) dispersions were synthesized with different defined Ce(NO$_3$)$_3$ content. All pristine dispersions were stable with different poly(tetramethylene oxide) glycol (PTMG) number average molecular weights (Mn) of 650, 1000, and 2000. The interaction between the carboxyl acid salt group and Ce(NO$_3$)$_3$ was analyzed by Fourier-transform infrared spectroscopy (FT-IR) and X-ray photoelectron spectroscopy (XPS) techniques. Coating hydrophilicity, water swelling (%), water contact angle, leaching, and corrosion protection efficiency were all affected when using different Ce(NO$_3$)$_3$ content and PTMG molecular weights. The maximal corrosion protection of the WBPU coating was recorded using a higher molecular weight of PTMG with 0.016 mole Ce(NO$_3$)$_3$ content.

Keywords: waterborne polyurethane; corrosion; cerium nitrate; coating

---

1. Introduction

Highly toxic inhibitors and solvents are being phased out and replaced by benign alternatives in coating industries [1,2]. In the last decade, it was established that volatile organic compounds (VOCs) and toxic inhibitors must be controlled to the lowest possible levels. The harmful impact of VOC coatings has led to the substitution of solvent-borne systems with waterborne systems [3]. Among the recently developed systems, polyurethane (PU) exhibits a special value over other waterborne coatings. Therefore, PU coatings are widely used in many industrial applications [4–7].

Using environmentally friendly polyurethane (PU) (e.g., waterborne polyurethane, WBPU) coatings for corrosion protection has garnered attention over the past couple of years, as certain pioneering researchers have improved the PU coating for barrier and scratch resistance [8–11]. PU is mainly composed of polyol (called the soft segment) and diisocyanate (called the hard segment). The polyol contributes to the elastomeric properties, and the diisocyanate controls the coating's mechanical properties. There are different types of polyols, such as polyether polyol, polyester polyol, silanol, etc. PU coating properties can be altered by tuning polyols and their contents [4,5]. In the last decade, PU technology has emerged as an attractive approach for developing next-generation PU coatings for broader industrial applications [4,5,11]. PU coatings are promising in antifouling, anticorrosion, antibacterial, anti-scratching and self-healing coatings [4,5,7]. Different anticorrosion coatings have been prepared by using different monomers, nanoparticles, and inhibitors [4,5].

Researchers have demonstrated that inhibitors are excellent corrosion protectors when used properly in coatings; however, their effectiveness also depends on the overall synthesis criteria, the interaction between the doped inhibitor and coating matrices, as well as the final deposited coating (thickness, adhesion, hydrophobicity, etc.) on the substrates [12–14]. Inhibitor solubility is another important criterion when using inhibitors in coatings. Usually, freely water-soluble inhibitors are less attractive than non-water-soluble inhibitors, since the former have a chance in the coating industry as those inhibitors have a chance of being easily leached, which decreases the life time of the dried coatings. However, the water-soluble inhibitors can be mixed easily with environmentally friendly water-based coatings. Therefore, water-soluble inhibitors might prove useful when the inhibitor can be retained in the coating for a longer time.

Different cerium salts (e.g., cerium phosphates, cerium cinnamates, and cerium silicates) exhibit good corrosion protection [15–17]. In some cases, water-soluble cerium salts such as cerium nitrate, $Ce(NO_3)_3$, and cerium chloride, $CeCl_3$, have also been used as inhibitors for the corrosion process [18]. In most of these cases, these inhibitors have been used in organic solvent-based coatings. It is difficult to find any reports on using cerium nitrate ($Ce(NO_3)_3$, a water-soluble inhibitor) in waterborne polyurethane (WBPU) coatings. In this study, three series of WBPU/$Ce(NO_3)_3$ dispersions were prepared with defined $Ce(NO_3)_3$ content using three different molecular weights of poly(tetramethylene oxide) glycol (PTMG) (M_n = 650, 1000, and 2000). The interaction of $Ce(NO_3)_3$ with PU was analyzed by Fourier-transform infrared spectroscopy (FT-IR) and X-ray photoelectron spectroscopy (XPS) analysis. The stability of the cerium salt dispersion was checked visually. The substrate for the coating was mild steel, and the synthesized WBPU/$Ce(NO_3)_3$ dispersions were used as a coating material. Corrosion testing was undertaken by potentiodynamic polarization (PDP) analysis. In addition, a leaching test at defined intervals was conducted to investigate the effect of $Ce(NO_3)_3$ content and PTMG molecular weight on the leaching of cerium salt.

2. Materials and Methods

Poly(tetramethylene oxide) glycol (PTMG, M_n = 650, 1000, 2000, Sigma Aldrich, St. Louis, MO, USA) samples were vacuum dried at 90 °C for three hours prior to use. Triethylamine (TEA, Sigma Aldrich), N-methyl-2-pyrrolidone (NMP, Sigma Aldrich), 4,4-dicyclohexylmethane diisocyanate (H_{12}MDI, Sigma Aldrich), and ethylene diamine (EDA, Sigma Aldrich) were used after dehydration with 4 Å molecular sieves for seven days. Dimethylolpropionic acid (DMPA, Sigma Aldrich), $Ce(NO_3)_3$ (Sigma Aldrich), and dibutyltindilaurate (DBTDL, Sigma Aldrich) were used as received.

2.1. Pristine Waterborne Polyurethane (WBPU) Dispersion Preparation

Pristine WBPU dispersion was prepared based on our previous report [7]. The solid content of the WBPU dispersion was fixed at 30 wt %. During preparation, the pre-polymer (NCO-terminated) was obtained by charging DMPA, H_{12}MDI, and PTMG. A very small amount of DBTDL was added during the reaction of PTMG and H_{12}MDI. 10 wt % methyl ethylketone (MEK) was added to the pre-polymer to lower the viscosity of pre-polymer. To neutralize the carboxyl group of DMPA, TEA was added to the pre-polymer mixture. Distilled water was added to the neutralized pre-polymer, which was followed by chain extension by adding EDA. Finally, MEK was evaporated from the WBPU dispersion.

2.2. Preparation of WBPU/$Ce(NO_3)_3$ Dispersion

A physical intermixing technique was applied to prepare the WBPU/$Ce(NO_3)_3$ dispersions. An exact amount (see Table 1) of $Ce(NO_3)_3$ was mixed with water to make a clear solution, which was mixed at the dispersion step of WBPU. The solid content was also fixed at 30 wt %.

Table 1. Sample designation, composition, and stability of dispersions. DMPA: dimethylolpropionic acid; EDA: ethylene diamine; $H_{12}MDI$: 4,4-dicyclohexylmethane diisocyanate; PTMG: poly(tetramethylene oxide) glycol; TEA: trimethylamine; WBPU: waterborne polyurethane.

Sample Designation	Composition (Mole)					PTMG (wt %)	Ce(NO$_3$)$_3$ (Mole)	Stability
	PTMG	DMPA	H$_{12}$MDI	TEA	EDA			
WBPU-650	0.525	0.760	1.83	0.760	0.550	32	0	Stable
WBPU-650-Ce-4	0.525	0.760	1.83	0.760	0.550	32	0.004	Stable
WBPU-650-Ce-8	0.525	0.760	1.83	0.760	0.550	32	0.008	Stable
WBPU-650-Ce-12	0.525	0.760	1.83	0.760	0.550	32	0.012	Unstable
WBPU-650-Ce-16	0.525	0.760	1.83	0.760	0.550	32	0.016	Unstable
WBPU-1000	0.525	0.760	1.83	0.760	0.550	43	0	Stable
WBPU-1000-Ce-4	0.525	0.760	1.83	0.760	0.550	43	0.004	Stable
WBPU-1000-Ce-8	0.525	0.760	1.83	0.760	0.550	43	0.008	Stable
WBPU-1000-Ce-12	0.525	0.760	1.83	0.760	0.550	43	0.012	Stable
WBPU-1000-Ce-16	0.525	0.760	1.83	0.760	0.550	43	0.016	Unstable
WBPU-2000	0.525	0.760	1.83	0.760	0.550	60	0	Stable
WBPU-2000-Ce-4	0.525	0.760	1.83	0.760	0.550	60	0.004	Stable
WBPU-2000-Ce-8	0.525	0.760	1.83	0.760	0.550	60	0.008	Stable
WBPU-2000-Ce-12	0.525	0.760	1.83	0.760	0.550	60	0.012	Stable
WBPU-2000-Ce-16	0.525	0.760	1.83	0.760	0.550	60	0.016	Stable
WBPU-2000-Ce-20	0.525	0.760	1.83	0.760	0.550	60	0.020	Unstable

The solid content of all samples was 30 wt %.

2.3. Coating onto Mild Steel Substrate

The synthesized coating solution was coated onto the mild steel through the autocoater. The wet coating thickness was 100 µm. All coatings dried at room temperature, and then oven dried at 70 °C for 24 h to remove the solvent perfectly.

2.4. Characterization

FT-IR spectroscopy (Impact 400D, Nicolet, Madison, WI, USA) was used to characterize the PU polymer.

The zeta potential value of the dispersion was analyzed by a Malvern Zetasizer 3000, zeta-potential analyzer (Malvern Instruments, Malvern, UK). During analysis, the temperature was fixed at 25 °C.

X-ray photoelectron spectroscopy (XPS) (ESCA 250 XPS, Thermo Scientific, East Grinstead, UK) was used to analyze the polymer surface.

For the swelling study, all films were immersed in water for 48 h at 30 °C and the swelling (%) was determined from the weight increase as [7]:

$$\text{Swelling (\%)} = (W - W_0/W_0) \times 100 \tag{1}$$

where W_0 is the weight of the dried film and W is the weight of the film at equilibrium swelling.

A Theta Optical tensiometer (Attension, Biolin Scientific, Helsinki, Finland) was used to analyze the water contact angle of the coatings.

The leaching rate was analyzed by UV (UV2600 UV-Vis spectrometer, Shimadzu, Kyoto, Japan) analysis, following a previous report [19]. The coated sample was immersed in water and slowly vibrated. After a certain time, the water was collected for UV analysis. From the spectra, the concentration of dissolved cerium ions was calculated.

A GAMRY3000 corrosion measurement system was used for potentiodynamic polarization (PDP) analysis, according to our previous report [20]. The reference electrode was −2 V vs. Ref. In this study, the electrochemical cell consisted of a prepared coated electrode as a working electrode, a graphite rod as a counter electrode, a saturated calomel electrode (SCE) as a reference electrode, and 3.5 wt % sodium chloride (NaCl) as the electrolyte. A surface Mask (GAMRY) of 1 cm^2 was used to mask the analyzed surface.

Potentiodynamic polarization plots were attained at the potential in the Tafel region with ±250 mV from open circuit potential (OCP) at a scan rate of 1 mV·s^{-1}. The corrosion rate (CR) of the coated samples was calculated from Tafel polarization curves using the following equation:

$$CR = 3268 \times i_{corr} \times EW/D \quad (2)$$

where i_{corr} represents the corrosion current density (mA·cm^{-2}), EW is the equivalent weight of the sample, and D is the density (g·cm^{-3}) of the sample. Further, the polarization resistance (R_p) was calculated using the following Stern–Geary equation:

$$R_p = \beta_a \times \beta_c / 2.303 \, (i_{corr}) \, (\beta_a + \beta_c) \quad (3)$$

where β_a and β_c represent the anodic and cathodic slopes, respectively. In addition, the inhibition efficiency (η) of coatings with and without the addition of Ce(NO$_3$)$_2$ were also calculated using the following relation:

$$\eta(\%) = R_p(\text{inhibit}) - R_p(\text{uninhibit}) / R_p(\text{inhibit}) \times 100 \quad (4)$$

where R_p(inhibit) and R_p(uninhibit) represent the polarization resistance of coatings with and without inhibitors, respectively.

3. Results and Discussion

The dispersion compositions and stabilities are summarized in Table 1. The major aspects considered during dispersion preparation were Ce(NO$_3$)$_3$ content and three different PTMG molecular weights. In all cases, the content of DMPA was fixed at 20 mole% [7].

First, in the polyol step, H$_{12}$MDI and DMPA were charged to make a NCO-terminated pre-polymer. In the second step, TEA was added to neutralize the carboxylic group. Water was added at the dispersion step. In the final step, EDA (with water) was added to the reaction mixture to complete the reaction.

In all cases, the peak at 2170 cm^{-1} disappeared (Figure 1) in the chain extension step; this confirmed the completion of the reaction [7]. The Ce(NO$_3$)$_3$ solution (mixed with water) was added drop-wise with vigorous stirring. The WBPU was identified by the presence of peaks at 1710 cm^{-1} and 3430 cm^{-1} for the C=O and N–H groups, respectively (Figure 1). Peaks were also recorded at 2795, 1540, and 1110 cm^{-1}. All these characteristic peaks confirmed the synthesis of PU.

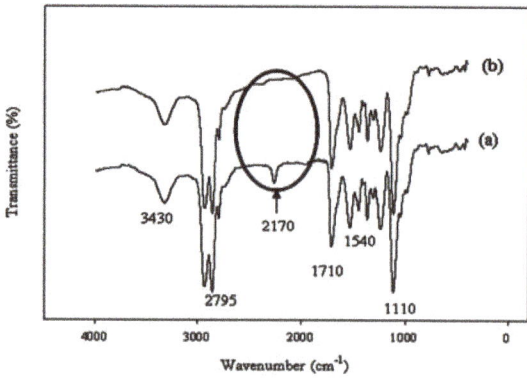

Figure 1. Fourier-transform infrared (FT-IR) spectra during polymerization (**a**) before chain extension; and (**b**) after chain extension.

The interaction between the Ce(NO$_3$)$_3$ and polyurethane was also studied using FT-IR analyses. Several new bands were observed (Figure 2). The bands at 1300 and 816 cm^{-1} were attributed to the NO$_3^-$ ion vibrations. The band at 530 cm^{-1} was likely to be associated with the Ce(III) group. To evaluate the proper interaction of Ce(NO$_3$)$_3$ with the carboxyl acid salt group and urethane/urea group, a curve fitting technique was applied from 1530 to 1770 cm^{-1} (Figure 3). It was found that by decreasing the molecular weight of PTMG, the value at 1643 cm^{-1} for the carboxyl group shifted slightly to a higher value. This implies a strong interaction between the carboxyl acid salt and Ce(NO$_3$)$_3$. However, it was not possible to measure the number (one, two, or three) of carboxyl acid groups that interacted with one Ce(III) from the FT-IR analysis.

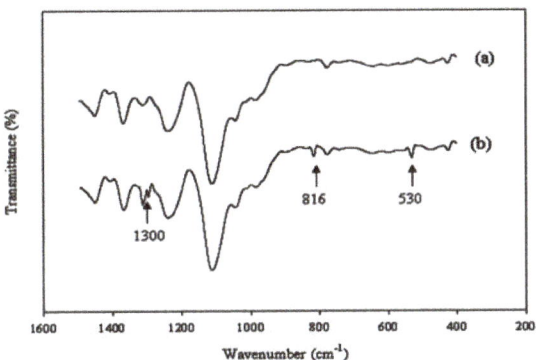

Figure 2. FT-IR spectra of WBPU during polymerization (a) without cerium salt; and (b) with cerium salt.

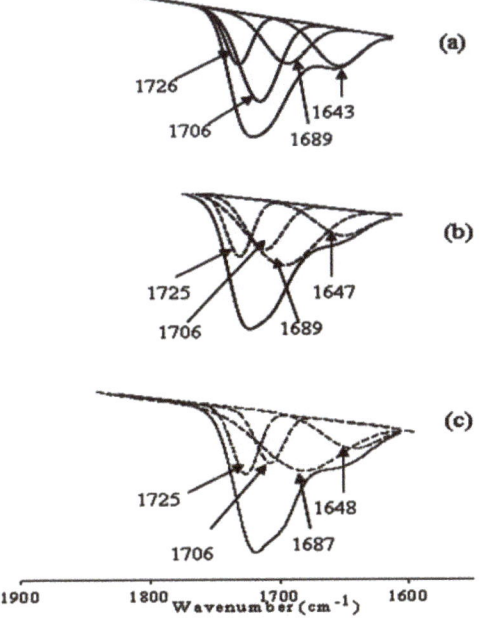

Figure 3. Curve fitting of FT-IR spectra of coatings with cerium salt of 0.008 mole from 1770 to 1530 cm^{-1}: (a) PTMG 2000; (b) PTMG 1000; and (c) PTMG 650.

The interaction between the carboxyl acid salt and Ce(NO$_3$)$_3$ was also analyzed through the XPS technique. A typical deconvoluted survey spectra using Ce(NO$_3$)$_3$ in the WBPU coating is shown in Figure 4. A peak at 887 eV appeared for the cerium(III) for all WBPU/Ce(NO$_3$)$_3$ coatings. However, with lower PTMG molecular weight (for M_n 1000 and 650), additional peaks also appeared at 888 eV for the cerium(III) of the WBPU/Ce(NO$_3$)$_3$ coatings. This implies that the interaction depended on the PTMG molecular weight. At higher molecular weights, the interacted carboxyl group might be one and showed one peak for cerium(III); whereas, at lower molecular weights, the interacted carboxyl group might be two or three and showed two/three peaks for cerium(III). The presence of a couple of peaks might also be due to the coordinated interaction between the cerium ion and carboxylate group. The C1s are classified into four component groups that correspond to the carbon atom of C=O (at 289.6–285.9 eV), C–C or C–H (at 282.2–285.8 eV), C–O (at 284.1–286.1 eV), and C–N (at 284.5–287.1 eV). A typical deconvoluted spectra for WBPU/Ce(NO$_3$)$_3$ is shown in Figure 5. Notably, the C=O is attributed to the DMPA and a broader peak from 290.7–286.0 (a higher value) with a recorded decrease of molecular weight (not shown). A broader peak usually comes from two or more peaks. This indicates that the interaction was two or three carboxyl acid salt groups. However, it was not possible to isolate these peaks from the curve analysis due to a very small broader peak. With increasing Ce(NO$_3$)$_3$, the intensity slightly increased for cerium(III) at respective values.

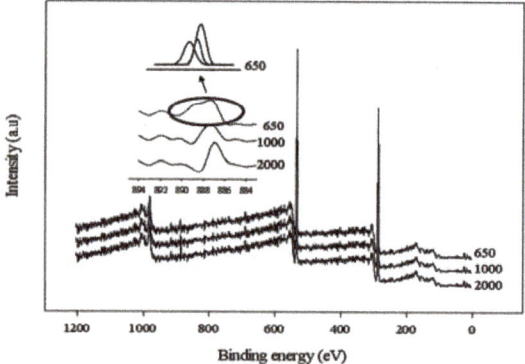

Figure 4. X-ray photoelectron spectroscopy (XPS) spectra of coatings with different molecular weight of PTMG with fixed cerium salt (0.08 mole).

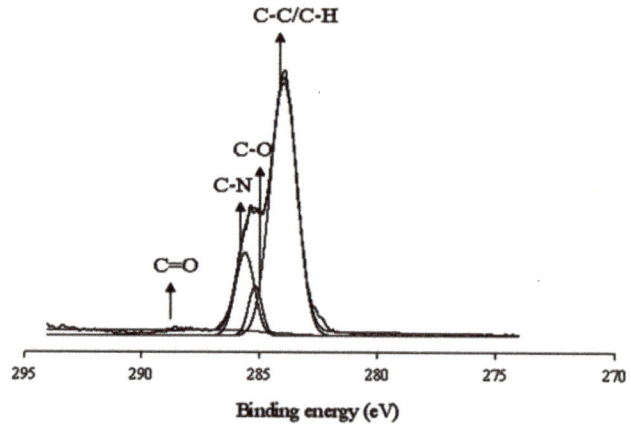

Figure 5. Deconvoluted XPS spectra of WBPU film.

All WBPU dispersions were stable with different PTMG molecular weights. The WBPU dispersion stability was affected by adding $Ce(NO_3)_3$. The dispersions were stable (no precipitation) up to a certain $Ce(NO_3)_3$ content, which also varied with the different PTMG molecular weights. The WBPU dispersions were stable up to 0.008 mole $Ce(NO_3)_3$ for all WBPU dispersions at the different PTMG molecular weights. Above 0.008 mole content, higher PTMG molecular weights yielded a higher capacity for loading $Ce(NO_3)_3$ in stable dispersions. Up to a 0.016 mole $Ce(NO_3)_3$ content could be used for a stable dispersion with PTMG 2000. This result clearly indicates that higher $Ce(NO_3)_3$ content can only be used with higher molecular weight PTMG. Researchers believe that the cerium salt (Ce^{3+}) complexes with mainly acid salts (in agreement with the FT-IR and XPS analysis), thus decreasing the free available carboxylic salt groups. As carboxyl salt is the key factor in producing a stable dispersion, increasing $Ce(NO_3)_3$ and carboxyl group complexation yielded unstable dispersions due a lack of sufficient free carboxyl salt groups. Scheme 1 shows the interaction of the cerium and carboxyl salt group complex. The precise basis for the higher cerium salt load with the higher molecular weight PTMG is unclear, and might be due to the higher content of PTMG, wherein the carboxyl groups are distal to cerium salt during complexation (see Scheme 1), which results in a lower chance of interaction of a single cerium and two or more carboxyl group complexation rate. Hence, the dispersion was stable with higher $Ce(NO_3)_3$ content and higher molecular weight PTMG. However, above 0.016 mole $Ce(NO_3)_3$, all WBPU dispersions were unstable. This might be due to the excess rate of interaction between the carboxyl acid salt and cerium.

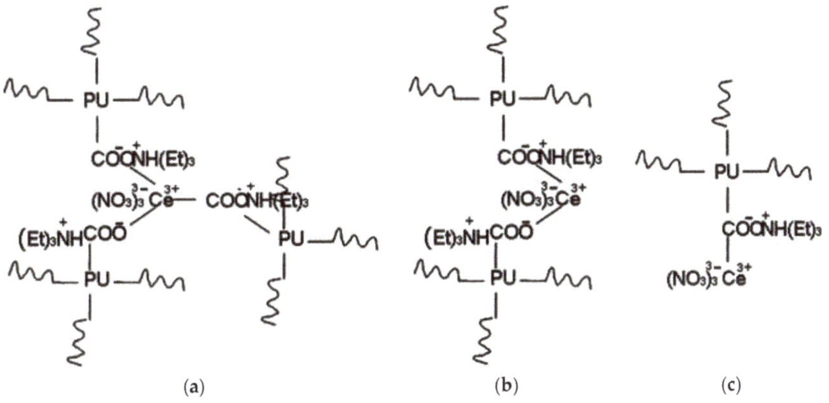

Scheme 1. Interaction of cerium salt with carboxyl acid salt: (a) three carboxyl acid salt; (b) two carboxyl acid salt; and (c) one carboxyl acid salt.

The dispersion stability can be described using zeta potential values. A higher magnitude of the value is an indicator of higher repulsive forces of particles, and predicts a long-term stable dispersion. If all particles in the dispersion exhibit a large negative or positive zeta potential, then they will tend to repel each other, and the particles will not tend to come together. Conversely, when the particles exhibit a lower magnitude of zeta potential, the particles come together easily to make the dispersion unstable (flocculation). Generally, the flocculation influence can be reduced or eliminated by increasing the particle charge. The zeta potential is summarized in Table 2.

Table 2. WBPU dispersions zeta potential and coatings water contact angle, water swelling, and leaching (%).

Sample	Zeta Potential (mV)	Water Contact Angle (°)	Water Swelling		Leaching (%)			
			24 h	48 h	12 h	24 h	36 h	48 h
WBPU-650	−0.49	58	17	18	–	–	–	–
WBPU-650-Ce-4	−0.38	58	20	22	50	65	71	73
WBPU-650-Ce-8	−0.22	55	23	23	63	72	77	80
WBPU-1000	−0.47	63	10	10	–	–	–	–
WBPU-1000-Ce-4	−0.42	63	10	10	35	47	52	55
WBPU-1000-Ce-8	−0.35	61	11	11	39	52	57	60
WBPU-1000-Ce-12	−0.23	58	13	13	46	59	63	69
WBPU-2000	−0.42	69	6	6	–	–	–	–
WBPU-2000-Ce-4	−0.40	68	6	6	9	19	24	30
WBPU-2000-Ce-8	−0.37	68	6	6	15	21	25	32
WBPU-2000-Ce-12	−0.35	67	7	7	17	23	28	35
WBPU-2000-Ce-16	−0.29	66	7	7	19	26	31	39

More negative values (below −40 mV) were recorded in pristine WBPU dispersions (without cerium salt). This implies a stable dispersion using different molecular weights of PTMG. However, the zeta potential value increased (less negative) by increasing the molecular weight of PTMG; this implies that the dispersion stability might have decreased with increasing PTMG molecular weight. With the addition of cerium salt, the zeta potential value changed for all dispersions. The value increased (e.g., less-stable dispersion) through an increase in $Ce(NO_3)_3$ content. However, the value changed dramatically in the low molecular weight PTMG-based dispersion. The addition of 0.004 mole cerium salt changed the zeta potential values to −11, −5, and −2 mV for WBPU-650-Ce-4, WBPU-1000-Ce-4, and WBPU-2000-Ce-4, respectively. The trend continued with increasing cerium salt content. The interaction between the carboxyl salt and cerium might be the reason why there were changes in the zeta potential value. In low molecular weight polyol, the interaction was strong due to the short chain polymer length. On the other hand, in high molecular weight polyol, the longer polymer chain might decrease the interaction due to stiff polymer chains.

Both the water contact angle and water swelling (%) tests were done to characterize the hydrophilicity of the coating. The results are summarized in Table 2. The coatings tended to swell with water at different rates. For water swelling, the WBPU films without cerium salt followed the order WBPU-2000 < WBPU-1000 < WBPU-650. As expected, the water contact angle exhibited the opposite order. These results imply that coating hydrophilicity decreased with an increase of molecular weight of PTMG. This was due to both the higher content of PTMG and the lower content of $H_{12}MDI$ in WBPU films. The same trend was also observed using cerium salt; with fixed cerium salt, lower hydrophilicity was observed in higher PTMG molecular weight-based film. At a fixed PTMG molecular weight, the water contact angle decreased and the water swelling increased with an increase in cerium salt content. This result was prominent using higher $Ce(NO_3)_3$ content. This implies that the interaction of $Ce(NO_3)_3$ with WBPU was strong enough to change the hydrophilicity of WBPU films when a certain amount of $Ce(NO_3)_3$ was present in the film. Especially after 24 h, the water swelling increased rapidly. This confirms that $Ce(NO_3)_3$ can have a detrimental effect with longer time applications for protection.

It was found that during leaching, the coatings with different content of $Ce(NO_3)_3$ formed a porous network that facilitated the transport of inhibitors from the coating. All recorded results are summarized in Table 2. At low inhibitor loading (except the 650 series), a very slow release rate was found, which might be due to the impermeability of the coating at a certain composition. At higher $Ce(NO_3)_3$ content, large amounts of $Ce(NO_3)_3$ leach out. This promotes rapid corrosion due to a large porous structure of the coating. The leaching rate was also different with different molecular weights of PTMG at a constant $Ce(NO_3)_3$ content. The low PTMG molecular weight-based coating showed a rapid leaching compared to the high PTMG molecular weight. A high hydrophilic character of the coating using a smaller molecular weight of PTMG made water penetration comparatively very easy, which effected faster leaching. Using PTMG at a higher molecular weight, the initial leaching was very slow. However, a moderate leaching rate was recorded after 24 h.

Earlier stage coating suitability was evaluated visually. We observed no mechanical damage for all coatings. The properly dried coatings were considered for corrosion tests. The PDP is a common technique used to investigate coating anticorrosion protection. Figures 6–8 show the typical PDP curves of the 650, 1000, and 2000 series for the WBPU coatings after immersion for 1 h, respectively.

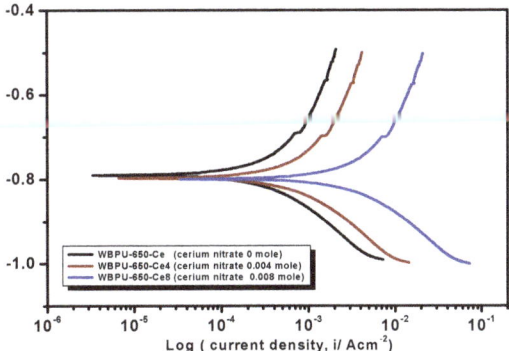

Figure 6. Potentiodynamic polarization (PDP) curves of the coatings of the PTMG 650 series with various cerium salt contents in 3.5% NaCl solution at a scan rate of $1~\text{mV} \cdot \text{s}^{-1}$.

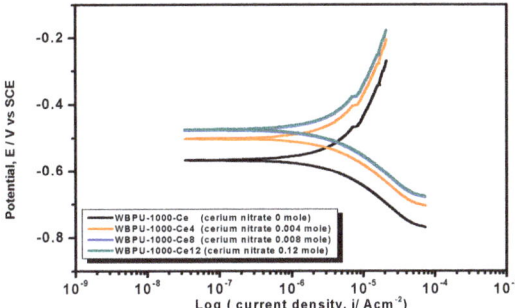

Figure 7. PDP curves of the coatings of the PTMG 1000 series with various cerium salt contents in 3.5% NaCl solution at a scan rate of $1~\text{mV} \cdot \text{s}^{-1}$.

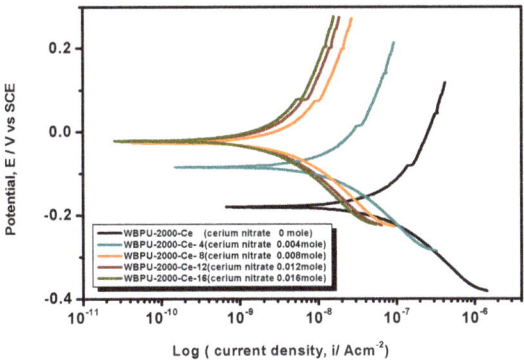

Figure 8. PDP curves of the coatings of the PTMG 2000 series with various cerium salt contents in 3.5% NaCl solution at a scan rate of $1~\text{mV} \cdot \text{s}^{-1}$.

The Tafel plot parameters are summarized in Tables 3 and 4. The E_{corr}, i_{corr}, R_p, $\eta\%$, and CR all varied with respect to PTMG molecular weight and Ce(NO$_3$)$_3$ content. The i_{corr} value was lower in higher PTMG when compared to those for lower with similar composition ratio of WBPU dispersion. For the i_{corr} value, the WBPU films without cerium salt followed the order WBPU-650 < WBPU-1000 < WBPU-2000. In all cases, a lower value of i_{corr} was recorded with the inclusion of Ce(NO$_3$)$_3$ when compared to the pristine WBPU coatings, respectively. This implies that the presence of Ce(NO$_3$)$_3$ increased the corrosion resistance. It was also recorded that a higher Ce(NO$_3$)$_3$ content resulted in a lower i_{corr} value for the coating; this also implies that corrosion protection efficiency increased with higher Ce(NO$_3$)$_3$ content. However, up to a certain Ce(NO$_3$)$_3$ content, the coating exhibited a lower i_{corr} value (higher corrosion resistance) in each series; then, the i_{corr} value very slightly decreased with increasing Ce(NO$_3$)$_3$ content. The coating WBPU-2000-Ce-16 showed the lowest i_{corr} value among all of the coatings; this implies the highest protection for metals. As this coating provided the maximum corrosion protection, the PDP was run at different intervals of 1 h, 24 h, and 48 h (Figure 9). The i_{corr} value dropped rapidly within 48 h. Though initial corrosion protection was very good, with time, the corrosion protection decreased rapidly. This implies that the coating corrosion resistance cannot be extended by only loading higher cerium salt content. As previous leaching tests confirmed a rapid leaching of the inhibitor within 24 h, the rapid corrosion resistance decrease might therefore be due to the leaching of the inhibitor. Though earlier corrosion protection was significant using Ce(NO$_3$)$_3$ in the WBPU coating, a rapid leaching of cerium salt made the coating less protective for a longer time in such conditions. Therefore, this coating can be used for protecting metals from atmospheric corrosion where the chance of the inhibitor leaching is much less due to the absence of an aqueous medium. To prolong the protection in wet conditions, the leaching should be controlled as much at a low rate by improving the coating barrier properties.

Table 3. Tafel parameters of coatings with different cerium nitrate content.

Sample	E_{corr} (mV)	i_{corr} (A·cm^{-2})	β_a (mV/dec.)	β_c (mV/dec.)
WBPU-650	−788	3.14 × 10^{-3}	85	71
WBPU-650-Ce-4	−784	6.27 × 10^{-4}	78	69
WBPU-650-Ce-8	−774	1.66 × 10^{-4}	62	74
WBPU-1000	−0.565	4.21 × 10^{-6}	85	67
WBPU-1000-Ce-4	−0.503	3.28 × 10^{-6}	71	88
WBPU-1000-Ce-8	−0.480	2.63 × 10^{-6}	92	72
WBPU-1000-Ce-12	−0.476	2.05 × 10^{-6}	76	63
WBPU-2000	−0.177	1.06 × 10^{-7}	81	93
WBPU-2000-Ce-4	−0.072	1.69 × 10^{-8}	64	86
WBPU-2000-Ce-8	−0.019	4.01 × 10^{-9}	74	82
WBPU-2000-Ce-12	−0.016	2.96 × 10^{-9}	69	76
WBPU-2000-Ce-16	−0.013	2.65 × 10^{-9}	74	90

Table 4. Polarization resistance (R_p), inhibition efficiency ($\eta\%$), and corrosion rate (CR) of coatings.

Sample	R_p (kΩ·cm^2)	$\eta\%$	CR (mm/Year)
WBPU-650	5.34	–	36.404
WBPU-650-Ce-4	25.36	78.94	7.269
WBPU-650-Ce-8	88.40	93.95	1.924
WBPU-1000	4067	–	4.88 × 10^{-2}
WBPU-1000-Ce-4	5206	21.87	3.80 × 10^{-2}
WBPU-1000-Ce-8	7360	44.74	3.04 × 10^{-2}
WBPU-1000-Ce-12	7980	49.03	2.37 × 10^{-2}
WBPU-2000	18,832	–	122.89 × 10^{-5}
WBPU-2000-Ce-4	1,060,800	98.22	19.59 × 10^{-5}
WBPU-2000-Ce-8	4,334,285	99.56	4.65 × 10^{-5}
WBPU-2000-Ce-12	5,307,692	99.64	3.43 × 10^{-5}
WBPU-2000-Ce-16	6,660,000	99.71	3.07 × 10^{-5}

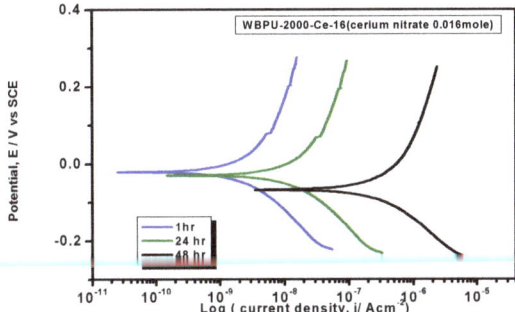

Figure 9. PDP curves of WBPU-2000-Ce-16 coating in different intervals in 3.5% NaCl solution at a scan rate of 1 mV·s^{-1}.

4. Conclusions

There is an urgent need to use fully green or less toxic materials in the coating industries. Unfortunately, most of the coatings use toxic solvents, inhibitors, and pigments. In this respect, the synthesized WBPU/Ce(NO$_3$)$_3$ dispersions can be considered as new green coatings. The number of interacted carboxyl groups played the key role in making the dispersion stable. The corrosion protection resistance increased with increasing Ce(NO$_3$)$_3$ content. The major challenge appeared in the presence of saline water, as a high release rate of cerium(III) was recorded in this condition. By improving the slow release rate of cerium(III), the protective properties of the coating can be significantly improved. Currently, this group is working on the controlled leaching of cerium(III) to improve the overall protection properties, which will be published in the near future.

Acknowledgments: This study was supported by the Center of Research Excellence in Corrosion (CoRE-C), King Fahd University of Petroleum and Minerals (KFUPM), Dhahran 31261, Saudi Arabia.

Author Contributions: Mohammad Mizanur Rahman and Md. Bashirul Haq conceived and designed the experiments; Mohammad Mizanur Rahman, Md. Hasan Zahir and Md. Bashirul Haq performed the experiments; Mohammad Mizanur Rahman, Md. Bashirul Haq, A. Madhan Kumar, and Dhafer A. Al Shehri contributed reagents/materials/analysis tools and characterization; and Mohammad Mizanur Rahman and Md. Bashirul Haq wrote the paper.

Conflicts of Interest: There is no potential conflict of interest.

References

1. Sungur, E.I.; Cotuk, A. Microbial corrosion of galvanized steel in a simulated recirculating cooling tower system. *Corros. Sci.* **2010**, *52*, 161–171. [CrossRef]
2. Syed, S. Atmospheric corrosion of carbon steel at marine sites in Saudi Arabia. *Mater. Corros.* **2010**, *61*, 238–244. [CrossRef]
3. Bat, E.; Gunduz, G.; Kisakurek, D.; Akhmedov, I.M. Synthesis and characterization of hyperbranched and air drying fatty acid based resins. *Prog. Org. Coat.* **2006**, *55*, 330–336. [CrossRef]
4. Chattopadhyay, D.K.; Raju, K.V.S.N. Structural engineering of polyurethane coatings for high performance applications. *Prog. Polym. Sci.* **2007**, *32*, 352–418. [CrossRef]
5. Chattopadhyay, D.K.; Webster, D.C. Thermal stability and flame retardancy of polyurethanes. *Prog. Polym. Sci.* **2009**, *34*, 1068–1133. [CrossRef]
6. Kim, B.S.; Kim, B.K. Enhancement of hydrolytic stability and adhesion of waterborne polyurethanes. *J. Appl. Polym. Sci.* **2005**, *97*, 1961–1969. [CrossRef]
7. Rahman, M.M.; Kim, H.D. Synthesis and characterization of waterborne polyurethane adhesives containing different amount of ionic groups (I). *J. Appl. Polym. Sci.* **2006**, *102*, 5684–5691. [CrossRef]

8. Christopher, G.; Kulandainathan, M.A.; Harichandran, G. Comparative study of effect of corrosion on mild steel with waterborne polyurethane dispersion containing graphene oxide versus carbon black nanocomposites. *Prog. Org. Coat.* **2015**, *89*, 199–211. [CrossRef]
9. Christopher, G.; Kulandainathan, M.A.; Harichandran, G. Highly dispersive waterborne polyurethane/ZnO nanocomposites for corrosion protection. *J. Coat. Technol. Res.* **2015**, *12*, 657–667. [CrossRef]
10. Coutinho, F.M.B.; Delpech, M.C. Synthesis and molecular weight determination of urethane-based anionomers. *Polym. Bull.* **1996**, *37*, 1–5. [CrossRef]
11. Coutinho, F.M.B.; Delpech, M.C.; Alves, A.S. Anionic waterborne polyurethane dispersions based on hydroxyl-terminated polybutadiene and poly(propylene glycol): Synthesis and characterization. *J. Appl. Polym. Sci.* **2001**, *80*, 566–572. [CrossRef]
12. Vakili, H.; Ramezanzadeh, B.; Amini, R. The corrosion performance and adhesion properties of the epoxy coating applied on the steel substrates treated by cerium-based conversion coatings. *Corros. Sci.* **2015**, *94*, 466–475. [CrossRef]
13. Rezaee, N.; Attar, M.M.; Ramezanzadeh, B. Studying corrosion performance, microstructure and adhesion properties of a room temperature zinc phosphate conversion coating containing Mn^{2+} on mild steel. *Surf. Coat. Technol.* **2013**, *236*, 361–367. [CrossRef]
14. Hao, Y.; Liu, F.; Han, E.; Anjum, S.; Xu, G. The mechanism of inhibition by zinc phosphate in an epoxy coating. *Corros. Sci.* **2013**, *69*, 77–86. [CrossRef]
15. Markley, T.A.; Forsyth, M.; Hughes, A.E. Corrosion protection of AA2024-T3 using rare earth diphenyl phosphates. *Electrochim. Acta* **2007**, *52*, 4024–4031. [CrossRef]
16. Shi, H.W.; Han, E.H.; Liu, F.C. Corrosion protection of aluminium alloy 2024-T3 in 0.05 M NaCl by cerium cinnamate. *Corros. Sci.* **2011**, *53*, 2374–2384. [CrossRef]
17. White, P.A.; Hughes, A.E.; Furman, S.A.; Sherman, N.; Corrigan, P.A.; Glenn, M.A.; Lau, D.; Hardin, S.G.; Harvey, T.G.; Mardel, J.; et al. High-throughput channel arrays for inhibitor testing: Proof of concept for AA2024-T3. *Corros. Sci.* **2009**, *51*, 2279–2290. [CrossRef]
18. Yu, M.; Liu, Y.; Liu, J.; Li, S.; Xue, B.; Zhang, Y.; Yin, X. Effects of cerium salts on corrosion behaviors of Si-Zr hybrid sol-gel coatings. *Chin. J. Aeronaut.* **2015**, *28*, 600–608. [CrossRef]
19. Soestbergen, M.V.; Baukh, V.; Erich, S.J.F.; Huinink, H.P.; Adan, O.C.G. Release of cerium dibutylphosphate corrosion inhibitors from highly filled epoxy coating systems. *Prog. Org. Coat.* **2014**, *77*, 1562–1568. [CrossRef]
20. Kumar, A.M.; Rahman, M.M.; Gasem, Z.M. A promising nanocomposite from CNTs and nano-ceria: Nanostructured fillers in polyurethane coatings for surface protection. *RSC Adv.* **2015**, *5*, 63537–63544. [CrossRef]

© 2018 by the authors. Licensee MDPI, Basel, Switzerland. This article is an open access article distributed under the terms and conditions of the Creative Commons Attribution (CC BY) license (http://creativecommons.org/licenses/by/4.0/).

Article

Poly(Phenylene Methylene): A Multifunctional Material for Thermally Stable, Hydrophobic, Fluorescent, Corrosion-Protective Coatings

Marco F. D'Elia [1], Andreas Braendle [1], Thomas B. Schweizer [1], Marco A. Ortenzi [2], Stefano P. M. Trasatti [3], Markus Niederberger [1] and Walter Caseri [1],*

[1] Department of Materials, ETH Zurich, 8093 Zürich, Switzerland; marco.delia@mat.ethz.ch (M.F.D.); andreas.braendle@mat.ethz.ch (A.B.); thomas.schweizer@mat.ethz.ch (T.B.S.); markus.niederberger@mat.ethz.ch (M.N.)
[2] Department of Chemistry, University of Milan, 20133 Milan, Italy; marco.ortenzi@unimi.it
[3] Department of Environmental Science and Policy, University of Milan, 20133 Milan, Italy; stefano.trasatti@unimi.it
* Correspondence: walter.caseri@mat.ethz.ch

Received: 29 June 2018; Accepted: 25 July 2018; Published: 7 August 2018

Abstract: Poly(phenylene methylene) (PPM) is a thermally stable, hydrophobic, fluorescent hydrocarbon polymer. PPM has been proposed earlier to be useful as a coating material but this polymer was isolated in relevant molar masses only recently, and in large quantities. Accordingly, the preparation of coatings based on PPM and their behavior was explored in this study, with the example of the metal alloy AA2024 as a common substrate for corrosion tests. Coatings free of bubbles and cracks were obtained by hot pressing and application of the following steps: Coating on AA2024 with a layer of polybenzylsiloxane to improve the adhesion between PPM and the metal surface, the addition of polybenzylsiloxane to PPM in order to enhance the viscosity of the molten PPM, and the addition of benzyl butyl phthalate as a plasticizer. Electrochemical corrosion tests showed good protection of the metal surface towards a NaCl solution, thanks to a passive-like behavior in a wide potential window and a very low current density. Remarkably, the PPM coating also exhibited self-healing towards localized attacks, which inhibits the propagation of corrosion.

Keywords: poly(phenylene methylene), aluminum alloy; AA2024; coatings by hot pressing; additives; fluorescence; corrosion protection

1. Introduction

Poly(phenylene methylene) (PPM) (Figure 1a) is a hydrocarbon polymer which consists of an alternating sequence of phenylene and methylene units. Therefore, PPM is structurally located between polyethylene and polyphenylene. Usually, the PPM products synthesized so far contain a mixture of *ortho-*, *meta-* and *para-*substituted phenylene units, with the latter reported to be dominant [1–3]. Interestingly, PPM unexpectedly exhibits photoluminescence, although the phenylene rings are not conjugated as in common fluorescent organic polymers [4]. This property was ascribed to homoconjugation, a rare phenomenon which can arise under certain geometrical conditions when conjugated π-orbital systems interact with each other, although they are separated by an electronically insulating group such as a methylene group [5,6], as illustrated for PPM in Figure 1b.

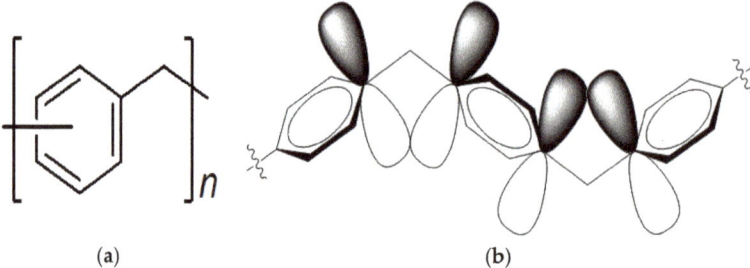

Figure 1. (a) Chemical structure of poly(phenylene methylene) (PPM); (b) Schematic representation of homoconjugation in PPM as a result of overlapping p-orbitals of phenylene rings which are separated by a methylene group.

Remarkably, in addition to hydrophobicity, PPM shows exceptionally high thermal stability [7–13] (onset of decomposition at 470 °C by thermogravimetric analysis under air atmosphere [4]). Further, PPM is structurally similar to phenol resins but possesses higher thermal stability [14]. Accordingly, PPM has been proposed as a valuable material for coatings [15–18]. In the panorama of protective coatings against corrosion of polymers with similar application methods and protection mechanism of PPM, there are epoxy resins, especially FBE formulations, and epoxy-polyaniline systems. These two classes of compounds present thermal stability of about 300 °C for epoxides, and about 200–240 °C for polyaniline [19,20]. Moreover, coatings with these polymers are not fluorescent and are also expected to be less hydrophobic than PPM, due to their chemical structures. In light of this, in addition to the high thermal stability of PPM (450–470 °C), PPM presents a unique combination of properties, such as fluorescence and hydrophobicity, where the latter can markedly influence the features of corrosion protective coatings.

However, coatings with PPM have not been investigated yet, presumably as PPM with adequate molar masses (number average molar mass M_n 10,000 g·mol^{-1}) has been isolated only recently [4,21]. Obviously, in the previously described synthesis procedures crucial conditions such as application of a steady nitrogen flow to remove the evolving hydrochloric acid and bulk polymerization under mechanical stirring were not adequately incorporated. Such parameters are even more important when polymerization proceeds via a step growth mechanism, as in the case of catalysts like SnCl$_4$. Notably, the synthesis of PPM is straightforward and it can readily be produced in large quantities (>100 g) on the laboratory scale.

Hence, in this work the preparation of coatings with PPM is explored. This includes the identification of a suited coating method and associated conditions for the maunfacture of coatings. Since it turned out in the course of the experiments that the film quality could be improved by additives, this aspect is also addressed. An aluminum alloy, AA2024, was selected as the substrate, pre-treated to enhance the coating adhesion, and used for corrosion tests. Accordingly, the ability of PPM in anti-corrosion protection is also examined in this study, revealing unusual self-healing properties.

2. Materials and Methods

Benzyl chloride (99%), tin(IV) chloride, phosphoric acid (85%), chloroform, and octaphenyl silsesquioxane (OP-POSS) was purchased from Sigma Aldrich (Buchs, Switzerland), benzyltriethoxysilane from Fluorochem (Hadfield, UK), and methanol (98%) from Merck (Darmstadt, Germany).

Poly(phenylene methylene) was synthesized by polymerization of benzyl chloride with SnCl$_4$ as a catalyst according to the literature [21], with number average molar mass M_n = 11,100 g·mol^{-1} and weight average molar mass M_w = 34,200 g·mol^{-1}, determined by gel permeation chromatography (GPC).

The common aluminum alloy AA2024 (also designated as Al-2024, containing 4.3%–4.5% copper, 0.5%–0.6% manganese, 1.3%–1.5% magnesium and less than 0.5% of other elements) was used as a substrate. Sheets of 12 cm in length, 3 cm in width and 4 mm in thickness were provided by Aviometal s.p.a (Varese, Italy). Samples of 4 cm in length were cut and subsequently polished with abrasive papers of 300, 500, 800 and 1200 grit. Immediately after polishing, the samples were cleaned by immersion in ethanol in an ultrasonic bath (Banderlin, Berlin, Germany) for 5 min. Then the samples were removed from the ethanol bath and the residual ethanol at the surface of the AA2024 samples was evaporated by means of a flush of nitrogen.

The freshly cleaned samples were covered with a layer of benzyltriethoxysilane by spin coating. Initially, 0.3 mL neat benzyltriethoxysilane was deposited in the center of the AA2024 substrates with a micropipette. Thereafter the spinning rate was increased from 0 to 3500 rpm. and maintained at this rate for 30 s. Finally, the samples were heated to 100 °C for 1 min, whereupon condensation of benzyltriethoxysilane to respective polysiloxanes commonly proceeds as described below. These samples were coated with PPM in absence and presence of siloxanes (OP-POSS and PBS, acronym of PBS see below) and plasticizer by hot pressing, as described in the section Results and Discussion, using about 100 mg of coating material.

A soluble condensation product of benzyltriethoxysilane, i.e., a polybenzylsiloxane designated as PBS, was synthesized under nitrogen atmosphere. A quantity of 0.40 mL (5.9 mmol) of phosphoric acid, 0.2 mL (10 mmol) of deionized water and 2.98 g (12 mmol) of benzyltriethoxysilane were transferred into a 100 mL three-neck flask equipped with a mechanical glass stirrer. The reaction was performed at room temperature, and after 10 min, a sol-gel reaction led to the formation of a highly viscous white material. After 24 h, during which evaporation of the reaction byproduct ethanol most likely took place under nitrogen flush, the material became an odourless white solid (2.43 g). A part, 1 g of the material, was dissolved in 5 mL of acetone, precipitated into 200 mL of water, filtered (cellulose filter) and dried under vacuum (10^{-2} mbar, 12 h) to obtain 0.78 g of PBS. ^{1}H-NMR spectra did not show anymore the characteristic –O–CH$_2$–CH$_3$ signal of ethoxy groups (triplet at 1.24 ppm, quadruplet at 3.83 ppm), i.e., the ethoxy groups initially present in benzyltriethoxysilane were virtually completely converted to ethanol under formation of siloxane units. Elemental analysis of the product (in % m/m): C 55.86, H 5.01.

Polymerization of benzyl chloride in presence of siloxanes was performed as follows: Under nitrogen atmosphere, 12 mL (104 mmol) of destabilized benzyl chloride was given to 0.132 g of additive (OP-POSS or PBS) in a 100 mL three-neck flask equipped with mechanical stirrer. After heating the mixture to 80 °C, 0.04 mL (0.34 mmol) of SnCl$_4$ was added. Polymerization was carried out under a constant nitrogen flow of 0.4–0.5 mL·min^{-1} to allow the produced HCl to escape from the reaction batch. After 3 h, the temperature was risen to 120 °C for 1 h and subsequently, due to the increase of viscosity upon polymerization, to 180 °C for 20 h. During the reaction, the color changed from deep red at the beginning of the reaction to a clear amber at the end of the reaction. The resulting reaction mixture was allowed to cool down to room temperature. An aliquot of 1 g of product was removed from the solids and dissolved in 5 mL of chloroform. This solution was poured into 250 mL methanol under stirring, filtered (cellulose filter) and dried under vacuum (10^{-2} mbar) for 12 h. The following quantities of product were obtained (the origin of the acronyms becomes evident in the section Results): 0.83 g of H-OP-POSS(P) and 0.74 g of H-PBS(P). The number average molar mass (M_n) of both polymers was in the range of 9000 to 11,000 g mol^{-1}.

For experiments with PPM comprising additives mixed in solution (PBS) or dispersion (OP-POSS), respectively, 3.00 g of PPM (synthesis according to the literature, see above) were stirred in 20 mL chloroform together with 0.03 g of siloxane (OP-POSS or PBS). The mixture was poured into 250 mL of methanol, and the precipitated material was filtered and dried at 10^{-2} mbar for 12 h. Yields for the products designated as H-OP-POSS(D): 2.96 g, H-PBS(D): 2.74 g.

Materials with plasticizer were produced starting with 200 mg (0.64 mmol) of benzyl butyl phthalate (BBP) which was added to 1.2 g of H-PBS(D) in 2 mL of tetrahydrofuran (THF) under magnetic stirring. When PPM and plasticizer were molten, the THF was evaporated at 60 °C under vacuum for 1 h and subsequently at room temperature under vacuum for 8 h. A quantity of 1.378 g product was obtained.

For rheology measurements, an Anton-Paar MCR-302 rheometer with parallel plates (Graz, Austria) was used, at a temperature of 90 °C. The values of complex viscosity, storage modulus and loss modulus were measured by collecting 20 points at angular frequencies starting from 400 to 0.4 s^{-1}.

Electrochemical characterizations were performed on AA2024 samples surface-modified with condensed benzyltriethoxysilane (see above) and a layer of H-PBS(D). The experiments were conducted in a simulated marine environment, i.e., at pH 6.9 ± 0.2 and 0.6 mol·L^{-1} NaCl (p.a., 99.8%, Sigma-Aldrich, Buchs, Switzerland) prepared with MilliQ water. The temperature was equal for all experiments (21.1 ± 1.0 °C).

A classic saturated calomel electrode (SCE = +242 mV vs. SHE) was used for all experiments. Potentiodynamic polarization curves were recorded after an initial delay time of 600 s for equilibration at open circuit potential (OCP), on a surface area of 1 cm^2 and at a rate of 0.167 mV·s^{-1}.

3. Results

3.1. Surface-Modified Metal Substrates

The high strength aluminum alloy AA2024 (also designated as Al-2024), widely employed in aerospace applications, was used as a substrate in this study. Due to its heterogeneous microstructure, this alloy is very sensitive to localised corrosion, particularly in chloride-containing environments [22]. These aluminum alloys were cleaned as described in the section Materials and Methods. When PPM films were prepared directly on cleaned AA2024, cracks arose in the final films within seconds after preparation, and the PPM readily detached from the surface (the same phenomenon occurred also when siloxanes were present in PPM as addressed below). In order to improve compatibility between the hydrophilic surface of the aluminum alloy and the hydrophobic PPM, the metal substrates were first covered by a layer of condensed benzyltriethoxysilanes of about 1 μm thickness (determined by SEM from cross sections). This strategy has been established already long ago for corresponding purposes [23,24]. As the Si–OEt (Et = ethyl) bonds in the ethoxysilanes are sensitive to water (including ambient humidity) it is generally assumed that highly crosslinked polysiloxanes (here polybenzylsiloxanes) are formed as hydrolysis products at the surfaces of the substrates under release of ethanol. Indeed, the polybenzylsiloxane layer prevented detachment of the PPM-based coatings applied as described below.

3.2. Coatings

Coatings of PPM were produced by pressing powders of PPM onto silane-pretreated metal substrates, with a polyetheretherketone (PEEK) foil to separate the PPM from the pressing instrument. Pressing was performed at 90 °C, i.e., above the glass transition temperature of the polymer (around 65 °C), for 30 s (no significant difference in the quality of the coatings was found by pressing for 60 s). The applied pressure amounted to 4 kPa. At lower pressures, the films were less uniform, and at higher pressures (up to 10 kPa) there was a tendency of enhanced formation of cracks in the final coatings. The thickness of the resulting films was about 40 μm (determined by SEM from cross sections and surface profilometry), and the coatings appeared very unniform and homogeneous

In the PPM films prepared by hot pressing, some cracks always arose upon cooling to room temperature even under optimum pressing conditions (Figure 2), and a few bubbles were usually found in the films, too. Deposition of PPM from solution (e.g., from chloroform or tetrahydrofuran) followed by solvent evaporation was also investigated but the films thus obtained contained much more bubbles, which could not be eliminated. It is known that an increase in the viscosity of molten polymers (as is the case with PPM in the hot press) can reduce bubble formation and prevent cracks [25–27]. The viscosity of PPM might be enhanced by well dispersed aromatic polynuclear siloxanes which are compatible with PPM. The aromatic units of these siloxanes might interact with the aromatic moieties of PPM and thus hamper the flow of the polymer chains. Therefore, the application of corresponding siloxanes was explored for the reduction of cracks and bubbles. These siloxanes were octaphenyl polysilsesquioxane (OP-POSS) and a polycondensation product of benzyltriethoxyxsilane (PBS). The OP-POSS molecules possess a core of rather cubic shape with 8 silicon atoms at the adges of the cube. Neighoring silicon atoms are connected by an oxygen atom, and a phenyl group is attached to each silicon atom. The employed OP-POSS was a commercial product, the other comopound was synthesized as described in the section Materials and Methods. The synthesis method was optimized to obtain virtually complete conversion of the ethoxy groups of the starting material and complete solubility of the condensation product in chloroform. Preliminary experiments revealed that a concentration of 1% m/m of siloxane was suited to improve the quality of PPM-based coatings.

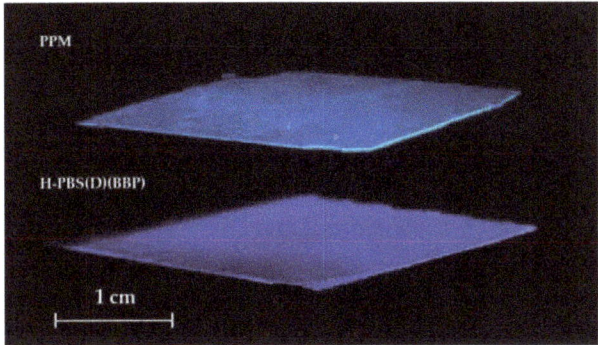

Figure 2. Top panel: PPM coating on polybenzylsiloxane-modified AA2024, showing cracks after the hot-pressed coating cooled down to room temperature. Bottom panel: PPM coating containing PBS and BBP (H-PBS(D)(BBP)), 2 weeks after hot-pressing. The pictures were taken irradiating with light having a wavelength of 365 nm, causing the characteristic blue fluorescence of PPM.

In order to incorporate the siloxanes, two methods were employed. On the one hand, the monomer benzyl chloride was mixed with the siloxanes (1% m/m) and polymerization was subsequently performed under the same conditions as for the synthesis of PPM itself. The siloxanes did not influence the course of polymerization significantly, in particular similar molar masses of PPM were obtained as in absence of siloxanes (M_n about 10,000 g mol^{-1}, see Materials and Methods). Also, the PPM with the siloxanes showed the common fluorescence of PPM (cf. Figure 2), and the glass transition temperature (57 °C, measured by DSC) was not significantly affected in all cases. The resulting material was also used directly for subsequent investigations without further work up. i.e., the mass ratio of additive and PPM necessarily remained in the order of the initial ratio of 1% m/m.

On the other hand, additives were introduced in PPM by mixing PPM solutions in chloroform with optically clear solutions of PBS and PBS-DS or opaque dispersions of OP-POSS. After precipitation in methanol, the resulting solids were filtered off. Although PPM as well as the siloxanes is insoluble in methanol, it cannot be excluded that the ratio of siloxane and PPM deviated from the initial ratio of 1% m/m. As a side note, preparation of films by evaporation of PPM solutions was generally not successful because upon solvent evaporation in the open atmosphere, a solid skin formed on top of the liquid (within about an hour in the case of chloroform). As a consequence, the solution underneath was trapped and the samples hardly dried (in the case of chloroform, they were not dry yet after a week), indicating very good barrier properties of PPM. Acceleration of evaporation by application of a vacuum lead to immense foaming.

Since the viscosity is considered to be an important factor for the preparation of crack-free films (see above), the complex viscosity and the damping factor (ratio of loss modulus and storage modulus) were determined. The hybrid-like materials containing PPM and PBS or OP-POSS are designated as H-PBS(D) or H-OP-POSS(D) when prepared from dispersion or solution and H-PBS(P) or H-OP-POSS(P) when prepared via in-situ polymerization of benzyl chloride. The complex viscosities of these materials are plotted against angular frequency in Figure 3. For all materials containing siloxanes higher values of complex viscosity were obtained compared to pure PPM (the molar masses of PPM were similar in all cases, M_n ca. 10,000 g·mol^{-1}). A particularly large increase of the complex viscosity was observed for the material H-PBS(D), compared to the polymer H-PBS(P) (increase by a factor 10–15, depending on the angular frequency) and PPM (increase by a factor 10–25). A retention of the shear thinning behavior is evident by the high increase of viscosity at higher angular frequencies (between 50–100 rad·s^{-1}), and plateau values were often evident at lower frequencies. In the case of materials obtained using OP-POSS the complex viscosity of the material prepared by the dispersion approach was below that prepared by the polymerization approach (as a remark, OP-POSS did not dissolve well in the solvent used to dissolve PPM for the preparation of H-OP-POSS(D)). The viscosity of H-OP-POSS(D), however, did not increase to the extent of that of H-PBS(D).

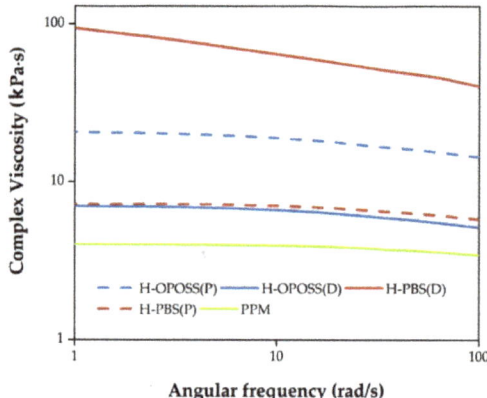

Figure 3. Complex viscosity of PPM without (green curve) and with additives (curves with other colors), as indicated in the figure.

Viscoelastic materials exhibit both elastic behavior and viscous behavior, respectively known in rheology as storage modulus and loss modulus. In order to determine how these parameters are balanced in PPM with siloxanes, values of damping factors were collected upon variation of the angular frequency in the same rheological experiment described above. In general, the addition of siloxane caused a decrease of the damping factor compared to pure PPM (Figure 4). With the exception of H-OP-POSS(D), however, the damping factors were above 1, implying the retention of the liquid-like structure in these cases. This indicates an absence of entanglements between siloxanes and PPM (as well as an absence of crosslinking but crosslinking is hardly possible with regard to the chemical nature of PPM and the siloxanes). According to the absence of chemical bonds between PPM and siloxanes these materials can be attributed to the so-called class 1 hybrid materials [28] that present high viscosity preserving a liquid-like structure in the molten phase, providing a suitable material for coating technologies. Yet for H-OP-POSS(D) the damping factor was far below 1, even though entanglements are highly improbable regarding the spheric structure of OP-POSS. Maybe segregation of OP-POSS in the polymer matrix can influence the rheologic measurement leading to these unexpected data. Notably, OP-POSS did not dissolve in chloroform but was present as an opaque dispersion upon mixing with the PPM solution, which implies the presence of OP-POSS aggregates. On the other hand, formation of a percolated network cannot be excluded.

The rather abrupt increase of the damping factor at frequencies below ca. 1 rad·s^{-1} (Figure 4) could essentially originate in a sol-gel-like transition. At high frequency the polymer forms a gel-like state or percolation network, while at low frequency the polymer is in a quasi-liquid state, i.e., the polymer molecules have sufficient time to move freely with respect to each other and can thus balance the external stress.

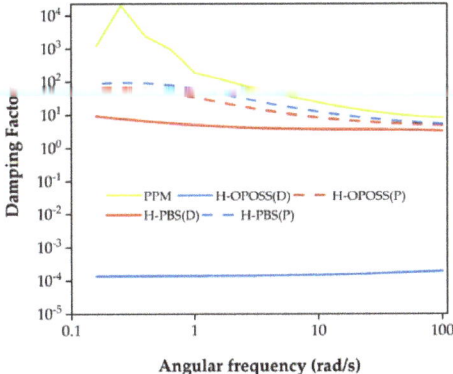

Figure 4. Damping factor of PPM without (green curve) and with additives (curves with other colors), as indicated in the figure.

Films with H-PBS(D) were prepared by hot pressing analogously to those described above with neat PPM. These materials were free of bubbles and formation of cracks was retarded (cracks emerged sometimes only after several hours or even after one whole day). In order to more efficiently prevent crack formation, a common plasticizer, benzyl butyl phthalate (BBP) was added to H-PBS(D), as the latter showed the highest viscosity (and still remained in a liquid-like condition). The addition of 17% m/m BBP showed good results. It was not possible to measure viscosities of this system at 90 °C as in absence of BBP (see above) because the presence of BBP made the molten polymer too low viscous for the available equipment. Nonetheless it was possible to prepare films with this system by hot pressing at the same conditions as for the films above. These coatings with BBP were free of bubbles and cracks, also over periods at least for months. Figure 2 shows a coating two weeks after hot pressing. Accordingly, corrosion tests were performed with such films, designated as H-PBS(D)(BBP). The plasticizer significantly decreased the glass transition temperature of the PPM from 65 to 35 °C.

3.3. Corrosion Tests

The electrochemical characterization of coatings of H-PBS(D)(BBP) on silane treated AA2024 was performed by cyclic polarization scans in order to obtain information on the resistance of the coatings to pit attacks. Also, potentiostatic analyses were performed for the observation of the resistance of the coatings over time. For this purpose, a fixed potential was applied on the cell for 28 h. The open circuit potential curves (OCP) showed that at equilibrium conditions, the potential curves of the coated samples fit the values of the OCP curve of a blank AA2024 substrate (in the range of −0.6 and −0.8 V related to SCE, in line with the literature [29]). The potential windows applied on the coated surfaces ranged from a −0.5 to 2.5 V (vs. SCE), and the corresponding current density was in the order of nanoampères, which confirms the barrier effect of the coating (Figure 5). This implies good anti-corrosion properties of the coating as the coating remained intact and the surface appeared uniform and compact, even when high potentials are reached. This fact is important as the efficiency of the coating relies on the separation of the metal surface from the corrosive environment. Thus, a physical barrier between metal surface and corrosive medium is established with the PPM-based coating.

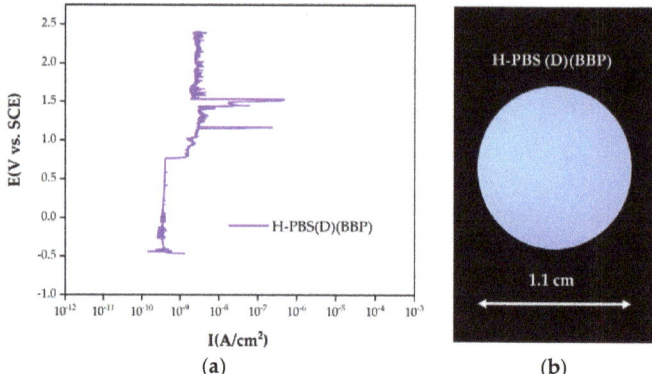

Figure 5. (a) Cyclic polarization curves of H-PBS(D)(BBP) coatings exposed to 0.6 mol·L^{-1} aqueous NaCl solution. (b) Surface of a coating exposed to the corrosion medium (picture taken at irradiation of light with wavelength of 365 nm, causing the characteristic blue fluorescence of PPM).

Testing the stability of anti-corrosion coatings over extended periods is important in order to evaluate the suitability for future applications. Therefore potentiostatic analysis was performed on coated surfaces, applying a constant potential at which the coating was stable according to the previous anodic polarization measurement, namely at about −0.3 V (cf. Figure 6). The results of this analysis shown in Figure 6 clearly demonstrate the resistance of the coating towards corrosion initiation over periods of at least 28 h. This stability is in the high-end of coatings investigated with this method. Remarkably, a very sharp peak occurred at ca. 5.5 h (Figure 6), after which the initial current intensity was re-established again. This shows that a metastable pit stopped rapidly (after a few minutes), thus demonstrating the ability of the H-PBS(D)(BBP) to self-regenerate under pit attacks, i.e., to stop the beginning of corrosion. Notably, the coatings looked still homogeneous after the potentiostatic experiments, in particular no evidence of propagating corrosion was visible.

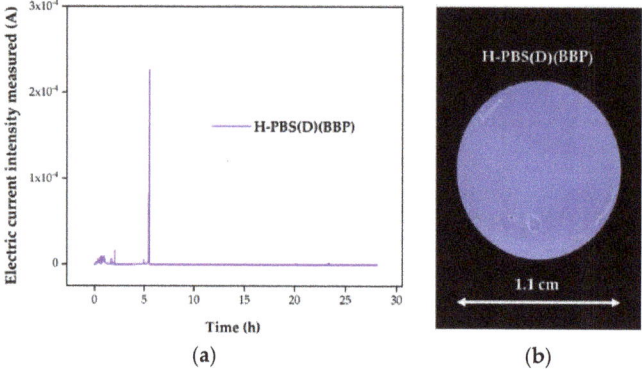

Figure 6. (a) Potentiostatic scan, at a constant potential of about −0.3 V, of H-PBS(D)(BBP)-coated substrates exposed to 0.6 mol·L^{-1} NaCl solution. (b) State of the H-PBS(D)(BBP) surface after the scan: Surface of a coating exposed to the corrosion medium for 28 h (picture taken at irradiation of light with wavelength of 365 nm, causing the characteristic blue fluorescence of PPM).

Corrosion inhibition may also be fostered by the pronounced hydrophobic nature of the coatings. The advancing contact angle of water on H-PBS(D)(BBP) layers amounted to 105° and the receding contact angle to 98°, i.e., the contact angle hysteresis is remarkably low (7°). Notably the contact angle hysteresis was even lower than that on AA2024 surfaces which amounted to 14°, according to the advancing and receding contact angles of 55° and 41°; respectively. The aluminum surfaces were not fully hydrophilic due to inevitable adsorption of atmospheric contaminants which proceeds within minutes [30,31]. As the contact angle hysteresis is an indication of surface homogeneity, it appears that the coated surfaces are even more homogeneous than the aluminum surface itself.

The electrochemical results of measured current densities, although limited to anodic polarization potentiodynamic analyses, allow to highlight a favorable protective behavior to corrosion by the PPM coating if compared to other organic coatings analyzed using the same electrochemical technique [32,33].

4. Discussion

Basically, poly(phenylene methylene) (PPM) is an attractive polymer for studies in the area of coatings since it is thermally stable, hydrophobic and fluorescent; the latter facilitates optical detection of inhomogeneities and cracks by observation under UV light. Moreover it can readily be prepared in large quantities (100 g and more) also on the laboratory scale, in a straightforward process. In spite of this, coatings of PPM have hitherto not been investigated, because the isolation of sufficiently pure PPM of considerable molar mass (10,000 g·mol^{-1}) was reported only recently. Therefore, the primary goal of this study was to explore the feasibility of PPM for formation of coatings, on the example of a metal substrate which is commonly used for corrosion tests (AA2024).

Several problems were identified in the preparation of coatings with PPM: Formation of bubbles, formation of cracks and detachment of PPM from the substrate. The latter was assumed to originate in the incompatibility of the highly apolar PPM and the highly polar surface of the AA2024 substrate. Already freshly polished AA2024 surfaces contain hydroxyl groups (evident from the symmetric and asymmetric Al–OH bending vibrations of boehmite detected in infrared spectra) [34]. Surface hydroxyl groups are commonly converted with trialkoxysilanes [23,24], RSi(OR')$_3$, in presence of some water which can also stem from ambient humidity. The trialkoxysilanes can react both with surface hydroxyl groups and water, thus forming a thin and highly crosslinked surface-bound polysiloxane layer under release of the alcohol HOR'. The organic group R is still present in that layer and should be compatible with the coating. Here R is a benzyl group which corresponds closely to the structure of the constitutional repeat units of PPM. Indeed, the respective *in-situ*-formed polysiloxane layers on AA2024 improved qualitatively adhesion of PPM to the substrate in an extent that detachment of PPM was not observed anymore.

The extent of bubble formation depended on the preparation method of the coating and the presence of additives. Hot pressing of PPM powders led to less bubbles in the coating than deposition of dissolved PPM followed by solvent evaporation. Gratifyingly, powder processing is also a more straightforward method than processing from solution because the dissolution step is not necessary in powder processing. In addition, powder processing is faster because in particular the evaporation of solvent in the solution process proceeds slowly.

Eventually, bubbles could be suppressed efficiently by addition of 1% *m/m* of PBS, an isolated polybenzylsiloxane, in coatings prepared by hot pressing, which was associated with a high increase in viscosity caused by the PBS. Accordingly, the interaction between the polymer chains is strong, thus apparently overcoming the energy required for bubble formation. PBS also retarded crack formation. This might be related to the stress accumulated to physical aging [35] during the cooling process after hot pressing. During ageing the free volume is reduced which is associated with a macroscopic volume contraction. This process can lead to widening of existing microcracks. It is common that such processes proceed within hours or days. Notably, fluorescence of PPM was preserved in presence of PBS, i.e., the PBS did not act as fluorescence quencher. In order to prevent cracks in a sustainable way, a plasticizer had to be added (BBP). In general, plasticizers enhance the elasticity of the system, and hence the system can adapt better to deformations which occur e.g., upon cooling the PPM-based coatings prepared by hot pressing. Viscosity is crucial to prevent bubbles, which can also lead to crack formation. In addition, crack formation can be caused by other factors such as internal stress inside the polymeric matrix. The latter can arise as a consequence of external stress occurring during film preparation (i.e., hot press) and the subsequent cooling process. A high elastic modulus and correspondingly a lower glass transition temperature (T_g) are helpful to attenuate those internal forces providing more mechanical stability to the coating. Thus, a compromise between the elastic modulus and the viscous modulus seems to be helpful in preventing bubbles and to mitigate internal force in the material.

Remarkably, electrochemical characterization revealed pronounced anti-corrosion properties of those PPM-based coatings, exhibiting a passive-like behavior in a wide potential window and a very low current density. This effect might be supported by the hydrophobic nature of the coating (advancing contact angle of water 105°), which may make the penetration of water to the metal substrate more difficult. Moreover, self-healing of the coating to pit attacks was found. Obviously, neat PPM is brittle below its glass transition temperature (about 65 °C). The stronger thermal shrinkage compared to the aluminum leads to tensile stresses, which cannot be taken by the PPM, due to missing mobility of the polymer chains. Since the plasticizer decreases the glass transition temperature, the chains regain their mobility and may thus close deficient locations, e.g., under the action of a pit attack.

With regard to future work, it could be explored if substituted PPMs show less of a tendency for crack formation, thus avoiding or at least reducing the need of additives for the generation of crack-free coatings. For instance, the synthesis of several methyl-substituted PPMs has already been reported [4,21] but they have not been investigated yet for coatings. Further, fluorinated PPMs might attract attention since in general they are more hydrophobic than their hydrogenated analogues. With a view to the alternatives addressed above, the presented study on PPM itself can be regarded as an initiation of more investigations on the class of poly(phenylene methylene)s in the area of coatings.

Author Contributions: M.F.D. and A.B. performed the synthesis of the materials and the manufacture of the coatings under supervision of W.C. and M.N.; T.B.S. designed and contributed to the rheologic measurements which were performed together with M.F.D.; the corrosion tests which were performed by M.F.D. under supervision of S.P.M.T. and M.A.O. regarding the macromolecular features of the materials, and the manuscript was written by W.C. and M.F.D.

Funding: This research was funded by the Swiss National Science Foundation (No. 200021_159719/1).

Acknowledgments: We thank Elena Tervoort for taking SEM images.

Conflicts of Interest: The authors declare no conflict of interest.

References

1. Blilncow, P.J.; Pritchard, G. Determination of the structure of polyethylarylmethylenes by ^{13}C nuclear magnetic resonance spectroscopy. *Polymer* **1987**, *28*, 1824–1828. [CrossRef]
2. Ul Hasasn, M.; Tsonis, C.P. Structural characterization of polybenzyls by high field ^{13}C-NMR spectroscopy. *J. Polym. Sci. A Polym. Chem.* **1984**, *22*, 1349–1355. [CrossRef]
3. Has, H.C.; Livingston, D.I.; Saunders, M. Polybenzyl type polymers. *J. Polym. Sci. A Gen. Pap.* **1955**, *15*, 503–514. [CrossRef]
4. Braendle, A.; Perevedentsev, A.; Cheetham, N.J.; Stavrinou, P.N.; Schachner, J.A.; Möosch-Zanetti, N.C.; Niederberger, M.; Caseri, W.R. Homoconjugation in poly(phenylene methylene)s: A case study of non-π-conjugated polymers with unexpected fluorescent properties. *J. Polym. Sci. B Polym. Phys.* **2017**, *55*, 707–720. [CrossRef]
5. Muller, P. Glossary of terms used in physical organic chemistry (IUPAC Recommendations 1994). *Pure Appl. Chem.* **1994**, *306*, 1077–1184. [CrossRef]
6. Ferguson, L.N.; Nnadi, J.C. Electronic interactions between nonconjugated groups. *J. Chem. Educ.* **1965**, *42*, 529–535. [CrossRef]

7. Baumberger, T.R.; Woolsey, N.F. Metal arene complexation of polybenzyl: Preparation and methylation. *J. Polym. Sci. A Polym. Chem.* **1992**, *30*, 1717–1723. [CrossRef]
8. Dreyer, D.R.; Jarvis, K.A.; Ferreira, P.J.; Bielawski, C.W. Graphite oxide as a dehydrative polymerization catalyst: A one-step synthesis of carbon-reinforced poly(phenylene methylene) composites. *Macromolecules* **2011**, *44*, 7659–7667. [CrossRef]
9. Grassie, N.; Meldrum, I.G. Friedel-crafts polymers—IX Later stages of the copolymerization of *p*-di(chloromethyl)benzene with aromatic substances. *Eur. Polym. J.* **1971**, *7*, 1253–1273. [CrossRef]
10. Gunes, D.; Yagci, Y.; Bicak, N. Synthesis of soluble poly(p-phenylene methylene) from tribenzylborate by acid-catalyzed polymerization. *Macromolecules* **2010**, *43*, 7993–7997. [CrossRef]
11. Arata, K.; Fukui, A. Toyoshima High catalytic activity of calcined iron sulphates for the polycondensation of benzyl chloride. *J. Chem. Soc. Chem. Commun.* **1978**, 121–122. [CrossRef]
12. Finocchiaro, P. Frazionamento di polimeri benzilici ottenuti a bassa temperature. *Boll. Sci. Fac. Chim. Ind. Bologna* **1968**, *26*, 255.
13. Hino, M.; Arata, K. Iron oxide as an effective catalyst for the polycondensation of benzyl chloride, the formation of para-substituted polybenzyl. *Chem. Lett.* **1979**, *8*, 1141–1144. [CrossRef]
14. Grassie, N.; Meldrum, I.G. Friedel-crafts polymers—2 Initial stages of the copolymerization of *p*-di(chloromethyl) benzene with benzene. *Eur. Polym. J.* **1969**, *5*, 195–209. [CrossRef]
15. Tsonis, C.P. Homogeneous catalytic polymerization of benzyl chloride leading to linear high molecular weight polymers: An elusive goal. *J. Mol. Catal.* **1990**, *57*, 313–323. [CrossRef]
16. Banks, R.E.; François, P.-Y.; Preston, P.N. Polymerization of benzyl alcohol in anhydrous hydrogen fluoride: An efficient synthesis of poly(phenylenemethylene). *Polymer* **1992**, *33*, 3974–3975. [CrossRef]
17. Klärner, C.; Greiner, A. Synthesis of polybenzyls by Suzuki Pd-catalyzed crosscoupling of boronic acids and benzyl bromides: Model reactions and polyreactions. *Macromol. Rapid Commun.* **1998**, *19*, 605–608. [CrossRef]
18. Som, A.; Ramakrishnan, S. Linear soluble polybenzyls. *J. Polym. Sci. A Polym. Chem.* **2003**, *41*, 2345–2353. [CrossRef]
19. Saliba, P.A.; Mansus, A.A.; Santos, D.B.; Mansur, H.S. Fusion-bonded epoxy composite coatings on chemically functionalized API steel surfaces for potential deep-water petroleum exploration. *Appl. Adhes. Sci.* **2015**, *3*, 22. [CrossRef]
20. Choi, Y.K.; Hyeong, J.K.; Sung, R.K.; Young, M.C.; Dong, J.A. Enhanced Thermal Stability of Polyaniline with Polymerizable Dopants. *Macromolecules* **2017**, *50*, 3164–3170. [CrossRef]
21. Braendle, A.; Schwendimann, P.; Niederberger, M.; Caseri, W.R. Synthesis and fractionation of poly(phenylene methylene). *J. Polym. Sci. A Polym. Chem.* **2018**, *56*, 309–318. [CrossRef]
22. Twite, R.L.; Bioerwagen, G.P. Review of alternatives to chromate for corrosion protection of aluminum aerospace alloys. *Prog. Org. Coat.* **1998**, *33*, 91–100. [CrossRef]
23. Witucki, G.L. A Silane Primer: Chemistry and applications of alkoxy silanes. *J. Coat. Technol.* **1993**, *65*, 57–60.
24. Plueddemann, E.P. Chemistry of Silane Coupling Agents. In *Silane Coupling Agents*, 2nd ed.; Plueddemann, E.P., Ed.; Springer: Boston, MA, USA, 1991; pp. 31–54.
25. Goldschmidt, P.D.A.; Streitberger, D.H.-J. *Basics of Coating Technology*; Vincentz: Hannover, Germany, 2007.
26. Gilleo, K.B. Rheology and Surface Chemistry. In *Coating Technology, Fundamentals, Testing and Processing Techniques*; Tracton, A.A., Ed.; Taylor and Francis: Boca Raton, FL, USA, 2007; pp. 1-1–1-12.
27. Chan, C.-M.; Venkatraman, S. Coating Rheology. In *Coating Technology, Fundamentals, Testing and Processing Techniques*; Tracton, A.A., Ed.; Taylor and Francis: Boca Raton, FL, USA, 2007; pp. 2-1–2-14.
28. Kickelbick, G. Introduction to hybrid materials. In *Hybrid Materials. Synthesis, Characterization, and Applications*; Kickelbick, G., Ed.; WILEY-VCH Verlag GmbH & Co. KGaA: Weinheim, Germany, 2007; pp. 1–46.
29. Trdan, U.; Grum, J. Evaluation of corrosion resistance of AA6082-T651 aluminium alloy after laser shockpeening by means of cyclic polarisation and EIS methods. *Corros. Sci.* **2012**, *59*, 324–333. [CrossRef]
30. Bain, C.D.; Troughton, E.B.; Tao, Y.-T.; Evall, J.; Whitesides, G.M.; Nuzzo, R.G. Formation of monolayer films by the spontaneous assembly of organic thiols from solution onto gold. *J. Am. Chem. Soc.* **1989**, *111*, 321–335. [CrossRef]

31. Troughton, E.B.; Bain, C.D.; Whitesides, G.M.; Nuzzo, R.G.; Allara, D.L.; Porter, M.D. Monolayer films prepared by the spontaneous self-assembly of symmetrical and unsymmetrical dialkyl sulfides from solution onto gold substrates: Structure, properties, and reactivity of constituent functional groups. *Langmuir* **1988**, *4*, 365–385. [CrossRef]
32. Talo, A.; Passiniemi, P.; Forsén, O.; Yläsaari, S. Polyaniline/Epoxy Coatings with Good Anti-Corrosion Properties. *Synth. Met.* **1997**, *85*, 1333–1334. [CrossRef]
33. Rodošek, M.; Rauter, A.; Perše, L.S.; Kek, D.M.; Vuk, A.S. Vibrational and corrosion properties of poly(dimethylsiloxane)-based protective coatings for AA 2024 modified with nanosized polyhedral oligomeric silsesquioxane. *Corros. Sci.* **2014**, *85*, 193–203. [CrossRef]
34. Kreta, A.; Rodošek, M.; Perše, L.S.; Orel, B.; Gaberšček, M.; Vuk, A.Š. In situ electrochemical AFM, ex situ IR reflection–absorption and confocal Raman studies of corrosion processes of AA 2024-T3. *Corros. Sci.* **2016**, *104*, 290–309. [CrossRef]
35. Jin, K.; Li, L.; Torkelson, J.M. Bulk physical aging behavior of cross-linked polystyrene compared to its linear precursor: Effects of cross-linking and aging temperature. *Polymer* **2017**, *115*, 197–203. [CrossRef]

 © 2018 by the authors. Licensee MDPI, Basel, Switzerland. This article is an open access article distributed under the terms and conditions of the Creative Commons Attribution (CC BY) license (http://creativecommons.org/licenses/by/4.0/).

Article

Anti-Corrosive and Scale Inhibiting Polymer-Based Functional Coating with Internal and External Regulation of TiO$_2$ Whiskers

Chijia Wang [1], Huaiyuan Wang [1,2,*], Yue Hu [1], Zhanjian Liu [1], Chongjiang Lv [1], Yanji Zhu [1] and Ningzhong Bao [3]

1. College of Chemistry and Chemical Engineering, Northeast Petroleum University, Daqing 163318, China; wangchijia@163.com (C.W.); zhaoyiming0109@163.com (Y.H.); liuzhanjian2012@163.com (Z.L.); Lvcj0316@126.com (C.L.); jsipt@163.com (Y.Z.)
2. School of Chemical Engineering and Technology, Tianjin University, Tianjin 300072, China
3. State Key Laboratory of Materials-Oriented Chemical Engineering, Nanjing Technology University, Nanjing 210009, China; 15201000195@163.com
* Correspondence: wanghyjiji@163.com; Tel./Fax: +86-459-650-3083

Received: 21 November 2017; Accepted: 2 January 2018; Published: 9 January 2018

Abstract: A novel multi-functional carrier of mesoporous titanium dioxide whiskers (TiO$_2$(w)) modified by ethylenediamine tetra (methylene phosphonic acid) (EDTMPA) and imidazoline was devised in epoxy coating to improve the anti-corrosion and scale inhibition properties of metal surface. Rigorous characterization using analytical techniques showed that a mesoporous structure was developed on the TiO$_2$(w). EDTMPA and imidazoline were successfully grafted on the outer and inner surfaces of mesoporous TiO$_2$(w) to synthesize iETiO$_2$(w). The results demonstrated that the corrosion resistance of the final iETiO$_2$(w) epoxy coating is 40 times higher than that of the conventional unmodified OTiO$_2$(w) epoxy coating. The enhanced corrosion resistance of the iETiO$_2$(w) functional coating is due to the chelation of the scaling cations by EDTMPA and electron sharing between imidazoline and Fe. Scale formation on the iETiO$_2$(w) coating is 35 times lower than that on the unmodified OTiO$_2$(w) epoxy coating. In addition, EDTMPA and imidazoline act synergistically in promoting the barrier property of mesoporous TiO$_2$(w) in epoxy coating. It is believed that this novel, simple, and inexpensive route for fabricating functional surface protective coatings on various metallic materials will have a wide range of practical applications.

Keywords: scale inhibition; anti-corrosion; mesoporous TiO$_2$ whiskers; organic coatings; mild steel

1. Introduction

Due to its high strength and ductility, steel is widely used in industrial and engineering structures. However, corrosion of steel often leads to degeneration in its properties, waste of resources, safety problems, and environmental issues [1,2]. Therefore, protecting steel from corrosion has become a topic of prime importance, especially to minimize economic losses. In this context, protective coatings are one of the most convenient and widely used methods for corrosion protection [3,4]. As the outermost layer on metallic structures, protective coatings provide physical shielding and anodic protection. Traditional methods for improving coating performances are mainly concentrated on increasing the thickness of the coating or increasing the content of active metal powers and new protective fillers. The abovementioned methods would result in a significant cost increase [5,6]. In order to reduce the costs and achieve functionally acceptable performance, a new generation of high-performance protective coatings are required, which can provide long-life anti-corrosion, as well as other functions [7].

Corrosion inhibitors, which can reduce or prevent corrosion reactions between a metal surface and its storage environment, are some of the most commonly used materials to enhance the corrosion resistance of metals [8]. Most organic corrosion inhibitors contain nitrogen, phosphorus, and sulfur heterocyclic compounds that can facilitate adsorption and film formation on metallic surfaces [9]. Imidazoline and ethylenediamine tetra (methylene phosphonic acid) (EDTMPA) have been demonstrated to be effective in inhibiting Fe corrosion [10,11]. Traditionally, corrosion inhibitors and scaling inhibitors are dispersed in the solution around the metal surface [12]. However, inhibitor molecules in the solution can move away rapidly in flow systems. Thus, the contact time between the inhibitors and the metal surfaces is shortened, leading to low protection efficiency and high cost. Therefore, incorporate inhibitors within the coating is an attractive proposition. Nonetheless, directly adding the small molecule organic matter into the coatings in a simple way will reduce their mechanical strength of coatings. Hence, new carriers must be developed to make full use of the inhibitors in protection against corrosion and scaling.

Micro/nanocontainers have been a subject of great scientific and industrial interest in the fields of medical science and materials science [13]. Within coatings, containers have been used as fillers, for carrying solid substrates, and for the encapsulation of liquid agents [14]. Among the various types of micro/nanocontainers available, microcapsules are considered attractive owing to their ability to uniformly disperse mutually incompatible fluids in each another [15]. However most microcapsules show poor mechanical properties compared to the traditional inorganic fillers, which constrains industrial use of microcapsules in coatings. In addition, the microcapsule shell should disintegrate to release the encapsulated agents [16]. Layer-by-layer assembly particle systems have received intense and growing attention in the past few years, but the complex techniques required have seriously compromised their scale-up manufacturing and applications [17]. Hence, it is necessary to develop physically and chemically stable porous carriers for the surface coating industry.

Titanium dioxide (TiO_2) is a widely used solid filler in the coating industry due to its ability to absorb ultraviolet light and improve the stability and weather resistance of coatings [18,19]. With the intent of taking advantage of these features, we developed mesoporous TiO_2 whiskers ($TiO_2(w)$) with a large surface area, high thermal stability, and good mechanical properties [20]. Compared to the traditional porous materials such as zeolite and aluminum oxide, mesoporous $TiO_2(w)$ is more suitable as a carrier in coatings. On the one hand, mesoporous $TiO_2(w)$ provides better acid-base resistance than aluminum oxide, so it can increase the resistance of coating to pitting corrosion performance in acid-base solution. On the other hand, the zeolite requires complex synthesis procedures and is costly.

In this study, a novel route was developed to fabricate functional epoxy coating containing with modified functional mesoporous $TiO_2(w)$ for steel substrates protection. The EDTMPA and imidazoline were modified on the surface of $TiO_2(w)$. Subsequently, the modified functional mesoporous $TiO_2(w)$ ($iETiO_2(w)$) carriers were dispersed in a silicone–epoxy resin coating. The nonwettability, surface roughness, anti-corrosion property, salt spray tests and the scale inhibition property of the prepared coating were investigated. It is expected that this work will pave a new way to design and fabricate functional epoxy coatings for industrial applications.

2. Materials and Methods

Detailed information on the materials and the characterization method is supplied in the Supplementary Materials.

2.1. Preparation of Mesoporous $TiO_2(w)$ Carriers

Mesoporous $TiO_2(w)$ carriers were prepared using potassium titanate, according to a previously described method [21,22]. K_2CO_3 and $TiO_2 \cdot nH_2O$ powders were mixed and sintered at 810 °C for 2 h to create a mixture with a TiO_2/K_2O molar value of 1.9. 10 g of this product was soaked in 7 mL of distilled water at room temperature in a closed container for about seven days. Later, the product was suspended in 100 mL of 0.1 M HCl under vigorous stirring for 10 h. In the final step, the resultant titanic acid suspension was calcined at a decomposition temperature of 500 °C to develop the mesoporous $TiO_2(w)$ carriers.

2.2. Internal and External Regulation of Mesoporous $TiO_2(w)$ Carriers

One gram of mesoporous $TiO_2(w)$ was added to 10 mL of H_2O_2 solution and the mixture was agitated for 3 h on a magnetic stirrer at room temperature for oxidation to occur. Later, 1 g of the oxidized mesoporous $TiO_2(w)$ ($OTiO_2(w)$) and 0.1 g of EDTMPA were added to a hydrothermal synthesis reactor (water injection rate 70%) and allowed to react at 80 °C for 24 h. At the end of this time period, the reaction mixture was filtered using 100 mL of deionized water. Subsequently, 1 g of the dried EDTMPA-grafted mesoporous $TiO_2(w)$ ($ETiO_2(w)$) and 0.1 g of imidazoline were added to 1 mL of deionized water in a beaker and aged for 12 h. After the aged mixture was dried at 80 °C for 12 h, the imidazoline dipped $ETiO_2(w)$ ($iETiO_2(w)$) was obtained.

2.3. Preparation of the Functional Epoxy Coatings

Eight grams of an organosilicon epoxy resin and 2 g of the carriers ($OTiO_2(w)$ or $ETiO_2(w)$ or $iETiO_2(w)$) were ultrasonically dispersed in 10 mL of ethyl acetate for 2 h. Subsequently, the coatings were prepared by spraying the ultrasonically dispersed solutions on as-treated steel plates (Q235, 80 mm × 80 mm × 1 mm) at a pressure of 0.6 MPa and curing at 180 °C for 2 h (manufacturers recommend). All the prepared coatings had an average thickness of 250 μm.

3. Results

In order to identify the composition of the synthesized functional carriers, Fourier transform infrared spectroscopy (FTIR) analysis of $OTiO_2(w)$, $ETiO_2(w)$, $iETiO_2(w)$, and imidazoline was carried out and the resultant spectra are shown in Figure 1a. $OTiO_2(w)$ exhibits four characteristic absorbance peaks, which are consistent with the reported functional groups on the surface of the as-received $TiO_2(w)$. The band at 3432 cm^{-1} can be assigned to the –OH stretching vibration [23]. The band observed at 937 cm^{-1} is characteristic of the C–P functional group. The peak at 1165 cm^{-1} is ascribed to the scissoring motion of P=O. The band at 1365 cm^{-1} is associated with the methyl group and the band at 1637 cm^{-1} is ascribed to C–H stretching [24]. Compared with the spectrum of $OTiO_2(w)$, the FTIR spectrum of $ETiO_2(w)$ powder displays a sharper peak of greater intensity at 3430 cm^{-1} due to the presence of hydroxyl groups in EDTMPA. These observations confirm the presence of EDTMPA on the surface of $ETiO_2(w)$. As we described in the experimental section, $ETiO_2(w)$ was filtered using 100 mL deionized water in order to remove any unreacted EDTMPA. Therefore, the FTIR results indicate that EDTMPA has been successfully grafted on the $OTiO_2(w)$ surface. Furthermore, no differences could be observed in the FTIR spectra of $iETiO_2(w)$ and $ETiO_2(w)$, which indicates that EDTMPA exists only on the surface of the $iETiO_2(w)$ carrier and the imidazoline in $iETiO_2(w)$ is proved to be infused into the pores of $TiO_2(w)$.

The textural properties of the $OTiO_2(w)$, $ETiO_2(w)$, and $iETiO_2(w)$ multi-functional carriers, including their Brunauer-Emmett-Teller (BET) surface areas, pore volumes, and pore diameters are summarized in Table 1. The BET surface area, total pore volume, and average pore diameter of $OTiO_2(w)$ are 52.37 m^2/g, 0.12 cm^3/g, and 9.2 nm, respectively. As reported elsewhere, the BET surface area of traditional TiO_2 systems, such as P25, is approximately 10 m^2/g [25]. The high BET surface area of $OTiO_2(w)$ is attributed to the whisker morphology and the numerous pores on the surface (as shown in Figure 2a). After the incorporation of EDTMPA, the S_{BET} of $OTiO_2(w)$ decreased slightly, indicating that EDTMPA is dispersed on the surface of $OTiO_2(w)$, which agrees well with the FTIR and X-ray diffraction (XRD) results. Moreover, upon being impregnated by imidazoline, the S_{BET} of $iETiO_2(w)$ decreased dramatically to 32.53 m^2/g. These results demonstrate that the internal pores of $TiO_2(w)$ were partly filled by imidazoline.

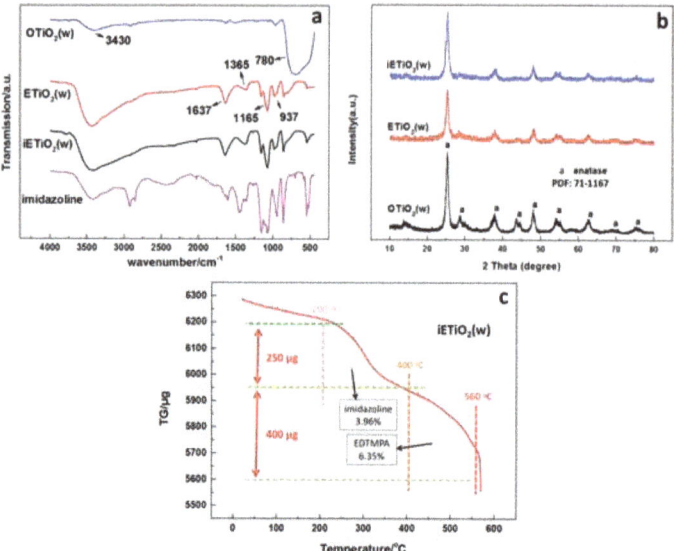

Figure 1. (a) FTIR spectra of OTiO$_2$(w), ETiO$_2$(w), iETiO$_2$(w), and imidazoline; (b) XRD patterns of OTiO$_2$(w), ETiO$_2$(w), and iETiO$_2$(w); (c) Thermogravimetric (TG) analysis curve of iETiO$_2$(w).

Table 1. Textural and structural properties of the multi-functional carriers.

Sample	S_{BET} (m^2/g)	V_p (cm^3/g)	Average Pore Diameter (nm)
OTiO$_2$(w)	52.37 ± 0.26	0.120 ± 0.006	9.2
ETiO$_2$(w)	48.74 ± 0.25	0.100 ± 0.005	8.2
iETiO$_2$(w)	32.53 ± 0.16	0.060 ± 0.003	7.4

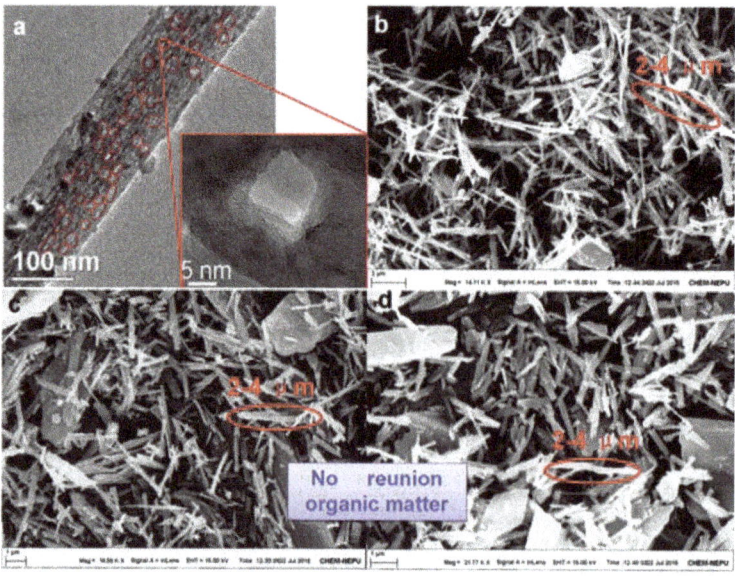

Figure 2. TEM image (a) of OTiO$_2$(w), SEM images of (b) OTiO$_2$(w), (c) ETiO$_2$(w), and (d) iETiO$_2$(w).

The thermal stability of iETiO$_2$(w) was studied using a thermogravimetry (TG) analyzer to confirm the existence of imidazoline. As shown in Figure 1c, there is a small weight loss at temperatures below 200 °C due to the volatilization of water. In the second thermal event between 200 °C and 400 °C, iETiO$_2$(w) loses about 250 μg (4% of its total mass) because of the oxidative decomposition of imidazoline [26]. Since imidazoline is absorbed into the pores on the whisker and linked to the inner surface of titanium dioxide via chemical bonds, it results in an increase in the oxidation decomposition temperature of imidazoline. The final thermal event in the TG curve is a representative of the EDTMPA decomposition diagram [27].

The crystalline phases in the multi-functional carriers were determined by XRD analysis. Figure 1b depicts the XRD patterns of OTiO$_2$(w), ETiO$_2$(w), and iETiO$_2$(w). The diffraction peaks of the anatase TiO$_2$ phase (PDF: 71-1167) can be identified in all the samples. The XRD results show that the TiO$_2$(w) crystal did not change after the oxidation and hydrothermal treatments. Moreover, apart from the peak related to anatase, no other peak could be observed in the three patterns, indicating that EDTMPA and imidazoline are evenly dispersed on the surface or inside the pores of OTiO$_2$(w). From the FTIR, XRD, TG, and BET results, it can be concluded that EDTMPA is spread on the outer surfaces of the mesoporous TiO$_2$(w) carriers while imidazoline is loaded in the internal pores of the mesoporous TiO$_2$(w) carriers.

In order to analyze the morphologies of the multi-functional carriers, scanning electron microscopy (SEM) was carried out. The SEM images of OTiO$_2$(w) and the modified mesoporous TiO$_2$(w) are displayed in Figure 2. The whisker shape of TiO$_2$(w) can be observed in Figure 2b; a large number of TiO$_2$ whiskers with an average diameter of 100 nm and an average length of 3 μm were evenly distributed throughout the sample. Interestingly, the transmission electron microscopy (TEM) image shows that there exist nano-sized pores (with an average diameter of 9.2 nm) on the surface of mesoporous TiO$_2$(w) (Figure 2a) and the representative pores have been marked. Figure 2c displays the surface morphology of ETiO$_2$(w). As shown in the image, there was no reunion of ETiO$_2$(w) with the organic matter and the morphology of the whisker remains unchanged even after the hydrothermal reaction. The morphology of the whiskers after immersion in imidazoline is shown in Figure 2d; the whiskers maintain their original morphological characteristics. The SEM results show that functional processing did not alter the pristine whisker morphology; EDTMPA and imidazoline are evenly dispersed on the exterior surface and the internal pores of the functional iETiO$_2$(w) carriers. These results are in good agreement with the FTIR observations.

The chemical modification mechanisms of imidazoline and EDTMPA on mesoporous TiO$_2$(w) are schematically illustrated in Figure 3. A large number of high-activity hydroxyl groups are generated on the surface of mesoporous TiO$_2$(w) after oxidation with H$_2$O$_2$. EDTMPA is a tetramethylene compound with four phosphate groups in its molecular structure. A hydroxyl group on the OTiO$_2$(w) surface and one of the phosphate groups of EDTMPA react under hydrothermal conditions. Its complex molecular structure and short chains make it difficult for EDTMPA to react with four hydroxyl groups at the same time. The phosphate groups that are not involved in the reaction serve as the scale inhibiting functional groups [28]. After the EDTMPA reaction, imidazoline groups are infused into the pores on the whiskers by immersion. The nitrogen in the imidazoline molecule has a lone pair of electrons in its outer shell. On the other hand, Ti^{4+} of mesoporous TiO$_2$(w) has an unsaturated 3d shell, which can accommodate two electrons [29]. Thus the corrosion inhibitor is deposited on the inner surface of the whisker under the dual action of physical and chemical adsorption. The results of specific surface characterization of different samples support the modification mechanisms described above.

Figures 4 and 5 present the EIS spectra of pure epoxy coating and the epoxy coatings containing OTiO$_2$(w) and iETiO$_2$(w), the spectra were obtained during long-term immersion conditions. The deterioration process of the pure epoxy coating can be divided into two stages (Figure 4a). During the first day, the coating exhibited a strong barrier effect, as indicated by the single large capacitive arc. The Nyquist plot of the spectra from seven to 60 days contained two time-constant semicircles, which are regarded as the capacitive loops, at medium and low frequencies. The medium frequency capacitive loops related to the charge transfer of corrosion reaction at the electrode surface. The loop at low frequencies is attributed to the charge transfer resistance (R_{ct}). These results indicate the permeation of oxygen, water, and corrosive ions (Cl$^-$) through the epoxy coating, finally resulting in under-film corrosion and coating delamination [30,31].

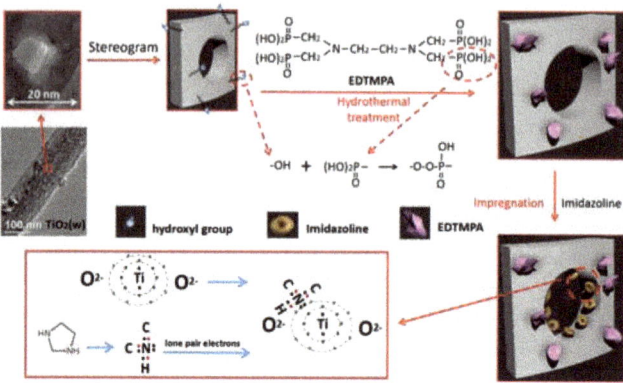

Figure 3. Modification of the carriers by EDTMPA and imidazoline.

The Nyquist plots (Figure 4b) constructed from the EIS spectra of the OTiO$_2$(w) epoxy coating were characterized by two large capacitive arcs from one to 60 days of immersion. The first arc corresponds to the capacitive impedance of the coating, which is measured by the diameter of the semicircles and the second arc corresponds to the polarization resistance process at the steel surface beneath the coating layer [32]. During the curing and application processes, many defects, such as micro-porosities, cavities, as well as free volumes are generated in the coating, resulting in the corrosive electrolyte penetrating into the coating matrix and leading to coating's degeneration and reduction of barrier performance [33].

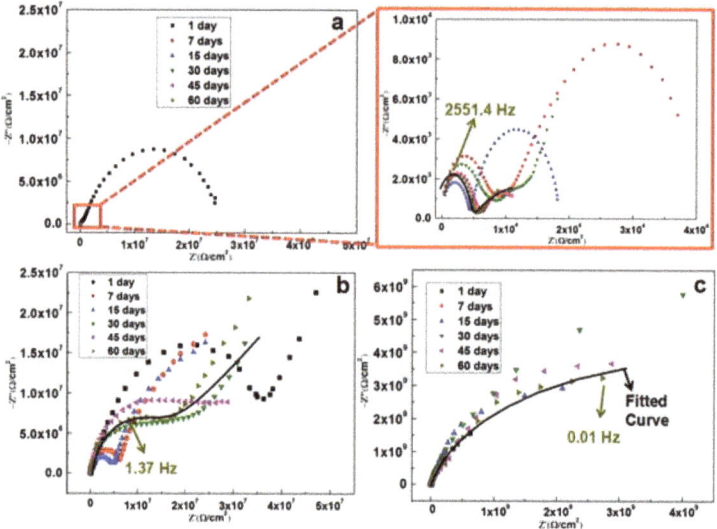

Figure 4. Nyquist plots of (**a**) pure epoxy coating, (**b**) OTiO$_2$(w) epoxy coating, (**c**) and iETiO$_2$(w) epoxy coating.

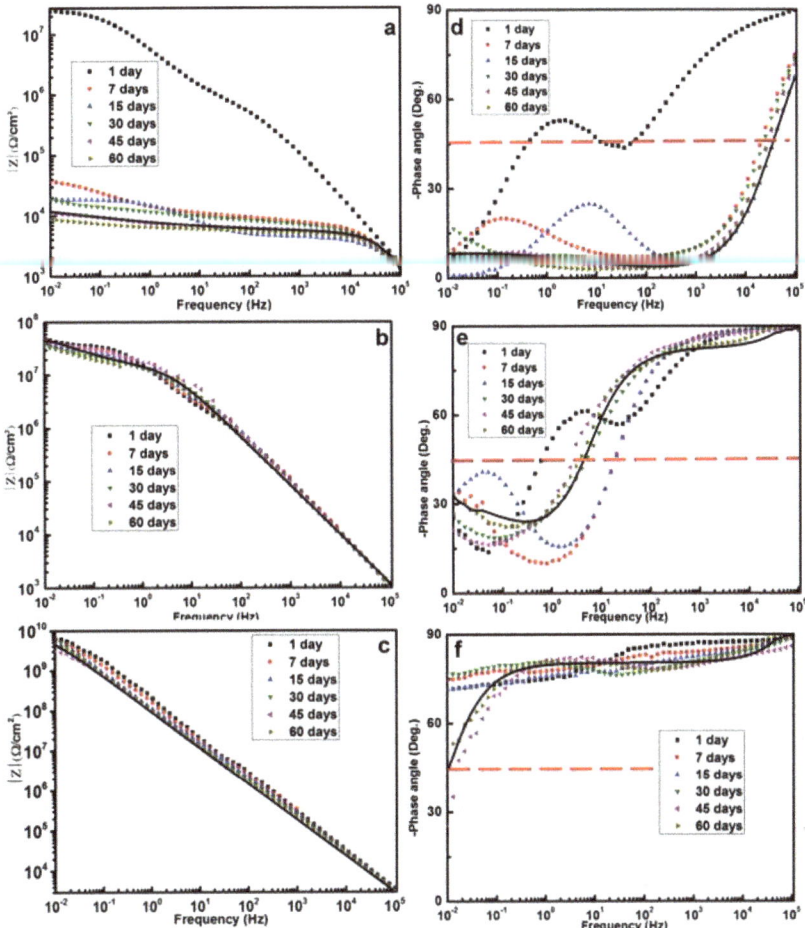

Figure 5. Bode plots of (a,d) pure epoxy coating, (b,e) OTiO$_2$(w) epoxy coating, and (c,f) iETiO$_2$(w) epoxy coating.

Furthermore, the long-term anti-corrosion performance of the iETiO$_2$(w) epoxy coating can be deciphered from Figure 4c. It can be seen that the diameter of the semicircle is about 8×10^8 Ω/cm^2 at day 1. Interestingly, as time goes by, the semicircle diameter starts to grow up to 10^9 Ω/cm^2, which indicates that the capacitive impedance of the coating can be increased by soaking it in a NaCl solution. At day 45, the semicircle diameter begins to reduce. This phenomenon is caused by the inhibitors activated by water molecules penetrate into the coating. Furthermore, the Nyquist plots for the iETiO$_2$(w) epoxy coating exhibit one semicircle over the whole frequency range during the 60 days exposure period, indicating a capacitive behavior and barrier type protection.

By using Bode plots, the barrier performance of the coatings was semi-quantitatively measured in terms of the impedance modulus at the lowest frequency ($|Z|_{0.01\text{ Hz}}$) [34]. In Figure 5a, the Bode plot for day 1 is a straight line, with a low-frequency impedance modulus that reaches a value of 10^7 Ω/cm^2. After seven or more days of immersion, overlapping straight lines can be observed with a low-frequency impedance modulus of ~10^5 Ω/cm^2, which is typical for epoxy coatings in harsh degeneration conditions [35]. The phase diagram confirms that there are already two time constants after one day of immersion, which might be associated with the electrochemical double layer capacitance on the solid/electrolyte interface. In Figure 5b, the $|Z|_{0.01\,Hz}$ values remain steady at $5 \times 10^7 \, \Omega/cm^2$ after 60 days of immersion. This result points out that $OTiO_2(w)$ can greatly improve the shielding effectiveness of the epoxy coating.

The breakpoint frequency (BF, frequency at 45° phase angle) values can also be obtained from the Bode diagrams. BF reflects the evolution of delamination and corrosion products beneath the coating. It can be observed from the phase diagram that the BF of pure epoxy is 100 Hz, while that of $OTiO_2(w)$ epoxy coating is at about 1 Hz. The BF value decreases with an increase in the modified depth of the coating. A lower plateau is seen at low frequencies, the phase diagrams are all characteristic with two time constants, and higher breakpoint frequencies were observed in Figure 5d–f, indicating a continuous decrease in the barrier properties during the immersion period for the pure epoxy coating and $OTiO_2(w)$ epoxy coating. Comparatively, the long-term anti-corrosion performance of the $iETiO_2(w)$ epoxy coating can also be reflected by the stable $|Z|_{0.01\,Hz}$ values and high phase angles (~70°) over a wide range of frequency during the 60 days' immersion. The high phase indicates the high resistance of the coating. The above BF results are found to be in good agreement with the Nyquist plots.

In the case of the $iETiO_2(w)$ epoxy coating, the $|Z|_{0.01\,Hz}$ values remained higher than $1 \times 10^9 \, \Omega/cm^2$ after 60 days of immersion, indicating that it had the highest shielding performance among all the tested coatings. The Nyquist plots (Figure 4c), the Bode plots (Figure 5c), and the phase diagrams (Figure 5f) of the $iETiO_2(w)$ epoxy coating indicate its pure capacitive behavior over the entire 60-day immersion period. The barrier properties remained constant despite the phase angles starting to decrease at lower frequencies at 45 and 60 days, which implies that a small amount of water penetrated into the coating [36]. The EIS results show that the barrier property of $iETiO_2(w)$ epoxy coating is 20–50 times higher than that of the $OTiO_2(w)$ epoxy coating. Water started penetrating into the $iETiO_2(w)$ epoxy coating 45 days later than it did in the case of the unmodified pure epoxy coating.

The electronic equivalent circuits (EECs) of the EIS results are displayed in Figure 6a–c. R_c and Q_c represent the resistance and constant phase element (CPE) of the coating, respectively. The charge transfer resistance (R_{ct}) and CPE of the electric double layer (Q_{dl}) appear after corrosion takes place beneath the coating. When the corrosion products diffuse through the pores in the coating, Warburg impedance (W) is added and serialized to R_{ct} [37,38]. The EEC shown in Figure 6a is used to fit the spectra of the $iETiO_2(w)$ epoxy coating from 0 to 60 days of immersion, during which no corrosion signals could be detected from underneath the coatings (Figure 7c). Figure 6b is related to the $OTiO_2(w)$ epoxy coating from one to 60 days of immersion (Figure 7b) and the pure epoxy coating during one day of immersion. It appears that the water molecules invaded the pure epoxy coating, leading to the corrosion of steel. In the case of a pure epoxy coating immersed for more than seven days, corrosion inducers diffused into the coating surface (Figure 7a). Therefore, Figure 6c conforms to the EEC of the pure epoxy coating from seven to 60 days of immersion. The fitting lines of the three coatings after immersion in 3.5 wt % NaCl solution for 60 days are shown in Figures 4 and 5. Results show that the black lines (fitting lines) are consistent with the trend of the dark green triangle, which means that the EECs are conforming to the circuit situation of each coating in a 3.5 wt % NaCl solution.

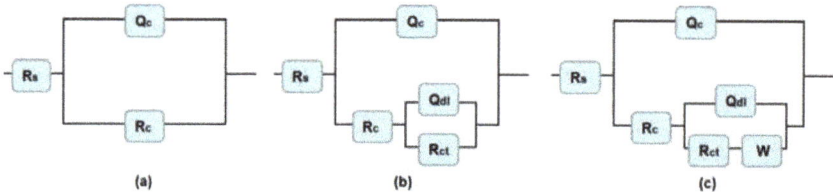

Figure 6. Electrical equivalent circuits used for fitting the EIS spectra. (**a**) one time constants equivalent circuits, (**b**,**c**) two time constants equivalent circuits.

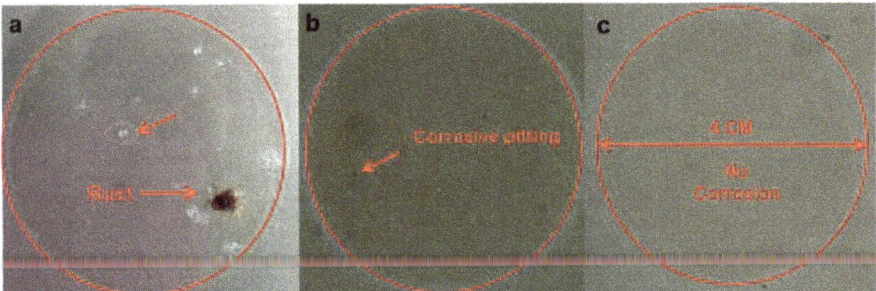

Figure 7. Surface topography of (**a**) pure epoxy coating, (**b**) OTiO$_2$(w) epoxy coating, and (**c**) iETiO$_2$(w) epoxy coating after ESI measurement.

Table 2 summarizes the results of the fitted parameters of the three coatings after immersion in 3.5 wt % NaCl solution for 60 days. It is evident that the R_t (R_t means $R_c + R_{ct}$) values of the pure epoxy coating are a thousand times smaller than those of the OTiO$_2$(w) epoxy coating. This observation indicates that the corrosion resistance of the pure epoxy coating deteriorated completely over 60 days. The R_t values of the OTiO$_2$(w) epoxy coating after 60 days of immersion are still very high. Interestingly, the R_t of the iETiO$_2$(w) epoxy coating is 40 times higher than that of the OTiO$_2$(w) epoxy coating. It should be noted that the CPE exponents (n_1) for all coatings at about 0.9. Q_c values are considered approximations of pure capacitances. The higher Rt and lower Q_c also contributed to the outstanding barrier property of the iETiO$_2$(w) epoxy coating. Results of the fitted parameters are in accordance with the Nyquist and Bode plots.

Table 2. Fitting parameters to simulate the EIS data of the pure epoxy coating, OTiO$_2$(w) epoxy coating, and iETiO$_2$(w) epoxy coating after immersion in 3.5 wt % NaCl solution for 60 days.

Coating	R_c	Q_c		R_{ct}	Q_{dl}	
	Ω cm^2	$Y_1 \times 10^{-9}$ (Ω^{-1} cm^{-2} s^n)	n_1	Ω cm^2	$Y_2 \times 10^{-6}$ (Ω^{-1} cm^{-2} s^n)	n_2
Pure epoxy coating	5531.8 ± 108	6.23 ± 0.15	0.86 ± 0.04	16,526 ± 400	201 ± 4	0.26 ± 0.01
OTiO$_2$(w) epoxy coating	1.19 × 10^6	3.86 ± 0.09	0.92 ± 0.05	1.20 × 10^8	0.076 ± 0.005	0.43 ± 0.02
iETiO$_2$(w) epoxy coating	8.96 × 10^9	1.91 ± 0.05	0.89 ± 0.05	–	–	–

It can be seen from the results described above that the impedance values of the pure epoxy coating and the OTiO$_2$(w) epoxy coating decreased with the increase of immersion time. This means that the corrosive electrolyte gradually diffused into the two coatings, while it is obvious that the inclusion of iETiO$_2$(w) had a significant impact on the epoxy coating's corrosion protection performance. The corrosion improving mechanism of the epoxy coating will be discussed in the following sections.

In order to further determine the long-term anti-corrosion properties of the coatings, accelerated corrosion tests were conducted in a neutral salt spray using a 5 wt % NaCl solution as the corrosive medium. Figure 8 shows the appearance of the four samples after 45 days of salt spray testing. Severe corrosion occurs on the pure epoxy coating. The brick red corrosion products are attributed to the formation of iron oxide. The red rust exists not only along the scratches but also spreads out to the unscratched coated surface. The above phenomenon is attributed to the poor protective and isolation performance of the pure epoxy coating. It can be seen from Figure 8b that the corrosion products exist mainly on the scratched area of the coating surface and no visible corrosion products can be seen on the unscratched surface. This indicates that the addition of OTTiO$_2$(w) can effectively prevent water molecules and chloride ions from penetrating through the coating, thus improving the protective performance of the epoxy coating. Different from the above results, it is evident from

Figure 8c that the iETiO$_2$(w) epoxy coating exhibits outstanding corrosion resistance during the whole salt spray test. No obvious brick red corrosion products are formed on the scratches or the other outer surface. The results show that the corrosion inhibitor encapsulated in the multi-functional carrier is released when water molecules penetrate into the coating. Figure 8d shows the salt spray test results of a commercial anti-corrosion coating (Rust Bullet) used as the reference sample. It can be understood from the salt spraying test results that the adhesion and corrosion resistance of the iETiO$_2$(w) epoxy coating are much higher than those of the commercial coating.

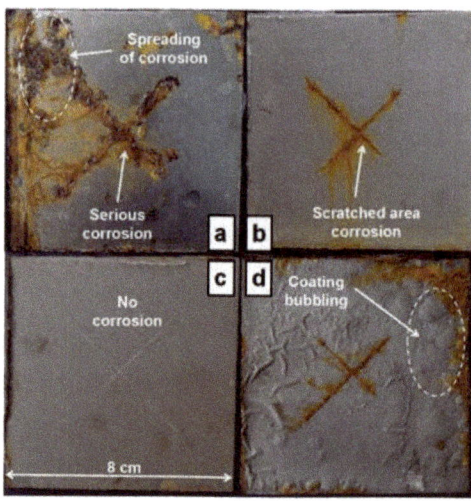

Figure 8. Salt spray tests results of (**a**) pure epoxy coating, (**b**) OTiO$_2$(w) epoxy coating, (**c**) iETiO$_2$(w) epoxy coating, (**d**) a commercial anti-corrosion coating (Rust Bullet).

In addition, electrochemical measurement for the defective iETiO$_2$(w) epoxy coating and the defected epoxy coating after 24 h salt spray tests is shown in Figure S1. The Nyquist plots (Figure S1a) constructed from the EIS data for the two defected coating were characterized by one large capacitive arc after 24 h salt spray tests. Both of the Bode plots exhibit a slight decline in the low-frequency range (Figure S1b). The phase diagram shows that there are already two evident time constants at 24 h of salt spraying, which confirms that the steel plates are exposed to the NaCl solution (Figure S1c). Furthermore, the capacitive arcs visible in the Nyquist plot and the value of $|Z|_{0.01\ Hz}$ (as shown in Figure S1) indicates that the charge transfer between the metal and the solution is hindered, which means that the iETiO$_2$(w) epoxy coating has a higher anti-corrosion performance.

Figure 9 illustrates the surfaces of coatings after being immersed in a CaCl$_2$/NaHCO$_3$ solution for 72 h. It can be seen that the surfaces of the pure epoxy coating and OTiO$_2$(w) epoxy coating are covered with a large number of cube-like blocks with an average size of 8 μm (Figure 9a,c). XRD analysis of the surface of the pure epoxy coating (Figure 9c) reveals that the cube-like entities are mainly the products of CaCO$_3$ fouling. Interestingly, Figure 9d reveals that there is very little CaCO$_3$ scaling on the surface of the multi-functional iETiO$_2$(w) epoxy coating [39]. In order to measure the scaling rate of the samples, the SEM images are converted to black and white two-value pictures. The gray value of the fouling material in the binary image is 255 and that of the others are 0. Finally, the scaling rate is obtained by calculating the proportion of 255 in the data. The results show that the scaling rates of the pure epoxy coating, OTiO$_2$(w) epoxy coating, and the iETiO$_2$(w) epoxy coating are 43%, 41.5%, and 1.2%, respectively. The scale formation on the iETiO$_2$(w) epoxy coating is 35 times lower than that on the OTiO$_2$(w) epoxy coating. The good scale inhibition effect of the iETiO$_2$(w) epoxy coating is obvious from the SEM analysis and the scale formation test.

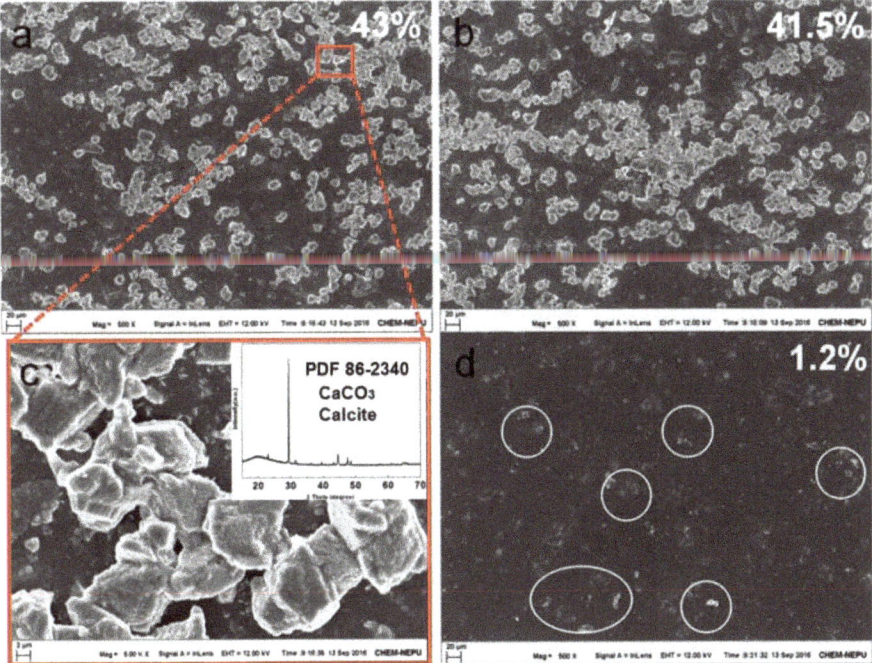

Figure 9. Scale inhibition property of (a,c) pure epoxy coating, (b) OTiO$_2$(w) epoxy coating, and (d) iETiO$_2$(w) epoxy coating.

The roughness data and pictures of the contact angle of pure epoxy coating, OTiO$_2$(w) epoxy coating, and iETiO$_2$(w) epoxy coating are shown in Figure 10. It can be seen from the pictures that the difference in roughness between the three samples is small. The pure epoxy coating is a smoother coating, with a contact angle at 98°. The roughness is improved and the contact angle is reduced to 80° when OTiO$_2$(w) is added to the coating. Although the roughness of the two epoxy coatings filled with TiO$_2$(w) is almost the same, the contact angle of the iETiO$_2$(w) epoxy coating is higher than the OTiO$_2$(w). The above phenomenon is mainly due to the –OH, which is hydrophilic, on the surface of OTiO$_2$(w). There are many fewer hydrophilic groups on the surface of iETiO$_2$(w) after surface modification. The improvement of the roughness of the coatings is mainly caused by adding fillers (TiO$_2$(w)) to the coating. The results above also demonstrated that the functional groups on the surface of the fillers can be exposed to the coating surface.

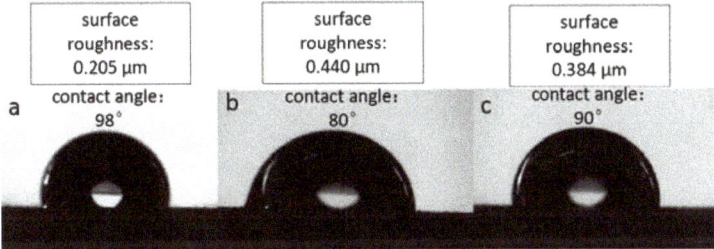

Figure 10. The roughness data and picture of contact angle of (a) pure epoxy coating, (b) OTiO$_2$(w) epoxy coating, and (c) iETiO$_2$(w) epoxy coating.

4. Discussion

Based on the chemical composition and characterization results described above, the anti-corrosion and scale inhibition mechanisms of the three coatings are discussed here. It has been shown in Figure 3 that the EDTMPA is a molecule containing four anti-scaling functional groups. When one phosphate group of EDTMPA is connected to one hydroxide radical on the surface of mesoporous TiO_2(w), the other unreacted anti-scaling functional groups are exposed to the outer surface, which can be used to bind cations. Then the scaling cations, such as Ca^{2+} and Mg^{2+}, will chelate with the anti-scaling functional groups, eventually leading to the formation of HCO_3^-, which cannot precipitate due to the lack of scaling cations (Figure 11) [40]. Meanwhile, there would always be EDTMPA exposed on the outer surface on the i$ETiO_2$(w) epoxy coating, even if the coating undergoes abrasion or breakage, since the functionalized i$ETiO_2$(w) is evenly dispersed in the epoxy coating.

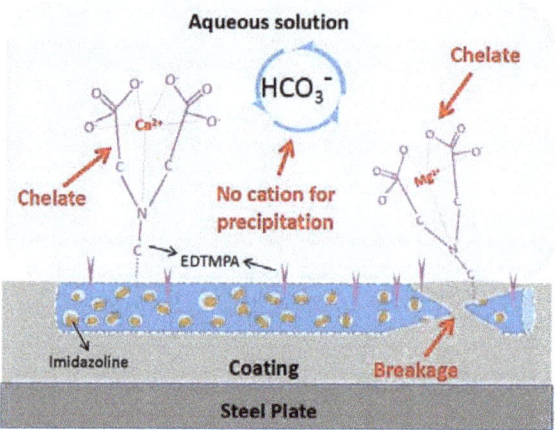

Figure 11. The scale inhibition mechanism of an i$ETiO_2$(w) epoxy coating.

The corrosion protection mechanism of the i$ETiO_2$(w) coating is displayed in Figure 12. It can be seen that the anti-corrosion process is divided into two steps. As displayed in the morphology section, mesoporous TiO_2(w) have an average diameter of 100 nm and an average length of 3 µm. When the water molecules permeate into the coating, mesoporous TiO_2(w) will obstruct the water molecules from directly penetrating into the coating/metal interface. After a long immersion or when the coating is scratched, water molecules reach the coating/metal interface (step 2); then imidazoline is activated by the water molecules and forms self-assembled monolayers on iron substrates, which also prevents rusting [41].

In summary, the barrier performance of the i$ETiO_2$(w) epoxy coating can be enhanced by the synergistic effect of EDTMPA and imidazoline. Compared to the traditional epoxy-based anti-corrosive coatings, the i$ETiO_2$(w) coating has the strongest anti-scaling and anti-corrosion properties. In consequence, using our particular experimental design, the processes of scale inhibition and corrosion inhibition can be mutually promoted.

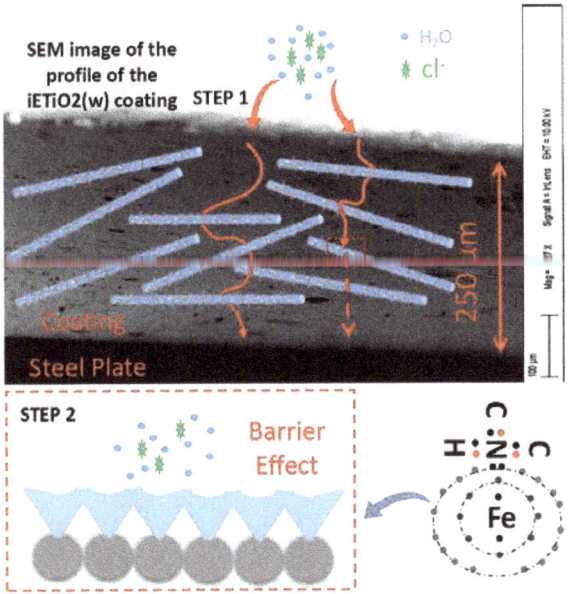

Figure 12. The corrosion protection mechanism of an iETiO$_2$(w) epoxy coating.

5. Conclusions

We have successfully fabricated a novel, multi-functional epoxy coating with outstanding scale and corrosion prevention properties for steel substrates by internally and externally regulating mesoporous TiO$_2$(w). The functional inhibitors imidazoline and EDTMPA modified the surface of mesoporous TiO$_2$(w), as evidenced by FTIR, TG, and BET analyses. The coatings were investigated for non-wettability, surface roughness, and anti-corrosion properties using salt spray tests and the scale inhibition. The main conclusions that could be drawn from our results are as follows:

- Analysis by electrochemical impedance spectroscopy showed that the resistance of the iETiO$_2$(w) epoxy coating exhibited outstanding barrier properties with a high resistance (8.96 × 10^9 Ω/cm^2) and long protection time, which indicates that the iETiO$_2$(w) epoxy coating exhibited excellent corrosion protection performance.
- Scale formation on the iETiO$_2$(w) epoxy coating was found to be 35 times lower than on the unmodified mesoporous TiO$_2$(w) epoxy coating, which means that the EDTMPA modified on the surface of the iETiO$_2$(w) plays an key role in the scale inhibition of the coating.

We believe that this novel route to fabricate anti-corrosion and scale-inhibiting coatings will inspire large-scale practical surface protection of structures such as steel pipelines, vessels, ships, and marine drilling platforms.

Supplementary Materials: The Supplementary Materials are available on http://www.mdpi.com/2079-6412/8/1/29/s1.

Acknowledgments: This research was financially supported by the National Young Top Talents Plan of China (2013042), the National Science Foundation of China (21676052, 21606042), the Northeast Petroleum University Innovation Foundation for Postgraduates (YJSCX2016-016NEPU), and the State Key Laboratory of Materials-Oriented Chemical Engineering (KL15-11).

Author Contributions: Chijia Wang and Huaiyuan Wang conceived and designed the experiments; Chijia Wang and Yue Hu performed the experiments; Zhanjian Liu, Chongjiang Lv, and Yanji Zhu analyzed the data; Chijia Wang, Huaiyuan Wang, and Ningzhong Bao wrote the paper.

Conflicts of Interest: The authors declare no conflict of interest.

References

1. Liu, T.; Yin, Y.; Chen, S.; Chang, X.; Cheng, S. Super-hydrophobic surfaces improve corrosion resistance of copper in seawater. *Electrochim. Acta* **2007**, *52*, 3709–3713. [CrossRef]
2. Merachtsaki, D.; Xidas, P.; Giannakoudakis, P.; Triantafyllidis, K.; Spathis, P. Corrosion protection of steel by epoxy-organoclay nanocomposite coatings. *Coatings* **2017**, *7*, 84. [CrossRef]
3. Barbhuiya, S.; Choudhury, M. Nanoscale characterization of glass flake filled vinyl ester anti-corrosion coatings. *Coatings* **2017**, *7*, 116. [CrossRef]
4. Wang, N.; Xiong, D.; Deng, Y.; Shi, Y.; Wang, K. Mechanically robust superhydrophobic steel surface with anti-icing, uv-durability, and corrosion resistance properties. *ACS Appl. Mater. Interfaces* **2015**, *7*, 6260–6272. [CrossRef] [PubMed]
5. Bai, N.; Li, Q.; Dong, H.; Tan, C.; Cai, P.; Xu, L. A versatile approach for preparing self-recovering superhydrophobic coatings. *Chem. Eng. J.* **2016**, *293*, 75–81. [CrossRef]
6. Banerjee, S.; Wehbi, M.; Manseri, A.; Mehdi, A.; Alaaeddine, A.; Hachem, A.; Ameduri, B. Poly (vinylidene fluoride) containing phosphonic acid as anticorrosion coating for steel. *ACS Appl. Mater. Interfaces* **2017**, *9*, 6433–6443. [CrossRef] [PubMed]
7. Liu, Y.; Liu, J. Design of multifunctional SiO_2–TiO_2 composite coating materials for outdoor sandstone conservation. *Ceram. Int.* **2016**, *42*, 13470–13475. [CrossRef]
8. Hu, K.; Zhuang, J.; Zheng, C.; Ma, Z.; Yan, L.; Gu, H.; Zeng, X.; Ding, J. Effect of novel cytosine-l-alanine derivative based corrosion inhibitor on steel surface in acidic solution. *J. Mol. Liq.* **2016**, *222*, 109–117. [CrossRef]
9. Kermannezhad, K.; Chermahini, A.N.; Momeni, M.M.; Rezaei, B. Application of amine-functionalized MCM-41 as PH-sensitive nano container for controlled release of 2-mercaptobenzoxazole corrosion inhibitor. *Chem. Eng. J.* **2016**, *306*, 849–857. [CrossRef]
10. Zhang, K.; Xu, B.; Yang, W.; Yin, X.; Liu, Y.; Chen, Y. Halogen-substituted imidazoline derivatives as corrosion inhibitors for mild steel in hydrochloric acid solution. *Corros. Sci.* **2015**, *90*, 284–295. [CrossRef]
11. Gao, Q.; Wang, S.; Luo, W.J.; Feng, Y.Q. Facile synthesis of magnetic mesoporous titania and its application in selective and rapid enrichment of phosphopeptides. *Mater. Lett.* **2013**, *107*, 202–205. [CrossRef]
12. Raja, P.B.; Sethuraman, M.G. Natural products as corrosion inhibitor for metals in corrosive media—A review. *Mater. Lett.* **2008**, *62*, 113–116. [CrossRef]
13. Cao, H.; He, J.; Deng, L.; Gao, X. Fabrication of cyclodextrin-functionalized superparamagnetic Fe_3O_4/amino-silane core–shell nanoparticles via layer-by-layer method. *Appl. Surf. Sci.* **2009**, *255*, 7974–7980. [CrossRef]
14. Shchukina, E.; Shchukin, D.; Grigoriev, D. Effect of inhibitor-loaded halloysites and mesoporous silica nanocontainers on corrosion protection of powder coatings. *Prog. Org. Coat.* **2017**, *102*, 60–65. [CrossRef]
15. Dong, B.; Wang, Y.; Fang, G.; Han, N.; Xing, F.; Lu, Y. Smart releasing behavior of a chemical self-healing microcapsule in the stimulated concrete pore solution. *Cem. Concr. Compos.* **2014**, *56*, 46–50. [CrossRef]
16. Kamburova, K.; Boshkova, N.; Boshkov, N.; Radeva, T. Design of polymeric core-shell nanocontainers impregnated with benzotriazole for active corrosion protection of galvanized steel. *Colloids Surf. A* **2016**, *499*, 24–30. [CrossRef]
17. Richardson, J.J.; Cui, J.; Björnmalm, M.; Braunger, J.A.; Ejima, H.; Caruso, F. Innovation in layer-by-layer assembly. *Chem. Rev.* **2016**, 14828–14867. [CrossRef] [PubMed]
18. Wang, H.; Zhang, S.; Wang, G.; Yang, S.; Zhu, Y. Tribological behaviors of hierarchical porous peek composites with mesoporous titanium oxide whisker. *Wear* **2013**, *297*, 736–741. [CrossRef]
19. Mazur, M.; Wojcieszak, D.; Kaczmarek, D.; Domaradzki, J.; Song, S.; Gibson, D.; Placido, F.; Mazur, P.; Kalisz, M.; Poniedzialek, A. Functional photocatalytically active and scratch resistant antireflective coating based on TiO_2 and SiO_2. *Appl. Surf. Sci.* **2016**, *380*, 165–171. [CrossRef]
20. Wang, H.; Cheng, X.; Xiao, B.; Wang, C.; Li, Z.; Zhu, Y. Surface carbon activated NiMo/TiO_2 catalyst towards highly efficient hydrodesulfurization reaction. *Catal. Surv. Asia* **2015**, *19*, 1–10. [CrossRef]
21. Xu, P.; Wang, R.; Ouyang, J.; Chen, B. A new strategy for TiO_2 whiskers mediated multi-mode cancer treatment. *Nanoscle Res. Lett.* **2015**, *10*, 94. [CrossRef] [PubMed]
22. He, M.; Lu, X.H.; Feng, X.; Yu, L.; Yang, Z.H. A simple approach to mesoporous fibrous titania from potassium dititanate. *Chem. Commun.* **2004**, *10*, 2202–2203. [CrossRef] [PubMed]

23. Wang, W.Y.; Zhao, X.F.; Ju, X.H.; Wang, Y.; Wang, L.; Li, S.P.; Li, X.D. Novel morphology change of AU-methotrexate conjugates: From nanochains to discrete nanoparticles. *Int. J. Pharm.* **2016**, *515*, 221–232. [CrossRef] [PubMed]
24. Wang, H.; Wang, C.; Bo, X.; Li, Z.; Jian, Z.; Zhu, Y.; Guo, X. The hydroxyapatite nanotube as a promoter to optimize the hds reaction of NiMo/TiO$_2$ catalyst. *Catal. Today* **2016**, *259*, 340–346. [CrossRef]
25. Shaw, S.S.; Sorbie, K.S. Synergistic properties of phosphonate and polymeric scale-inhibitor blends for barium sulfate scale inhibition. *SPE Prod. Oper.* **2015**, *30*, 16–25. [CrossRef]
26. Alla, E.M.A.; Abdel-Hamid, M.I. Kinetics and mechanism of the non-isothermal decomposition. Some Ni(ii)-carboxylate-imidazole ternary complexes. *J. Therm. Anal. Calorim.* **2000**, *62*, 769–780. [CrossRef]
27. Liu, H.; Cui, Y.; Li, P.; Zhou, Y.; Chen, Y.; Tang, Y.; Lu, T. Polyphosphonate induced coacervation of chitosan. Encapsulation of proteins/enzymes and their biosensing. *Anal. Chim. Acta* **2013**, *776*, 24–30. [CrossRef] [PubMed]
28. Oshani, F.; Marandi, R.; Rasouli, S.; Farhoud, M.K. Photocatalytic investigations of TiO$_2$–P25 nanocomposite thin films prepared by peroxotitanic acid modified sol–gel method. *Appl. Surf. Sci.* **2014**, *311*, 308–313. [CrossRef]
29. Bun, H.; Monjanel-Mouterde, S.; Noel, F.; Dur, A.; Cano, J.P. Inhibition properties of self-assembled corrosion inhibitor talloil diethylenetriamine imidazoline for mild steel corrosion in chloride solution saturated with carbon dioxide. *Corros. Sci.* **2013**, *77*, 265–272. [CrossRef]
30. Jie, H.; Xu, Q.; Wei, L.; Min, Y.L. Etching and heating treatment combined approach for superhydrophobic surface on brass substrates and the consequent corrosion resistance. *Corros. Sci.* **2015**, *102*, 251–258. [CrossRef]
31. Mahallati, S.; Saremi, E. An assessment on the mill scale effects on the electrochemical characteristics of steel bars in concrete under DC-polarization. *Cem. Concr. Res.* **2006**, *36*, 1324–1329. [CrossRef]
32. Pour-Ali, S.; Dehghanian, C.; Kosari, A. Corrosion protection of the reinforcing steels in chloride-laden concrete environment through epoxy/polyaniline–camphorsulfonate nanocomposite coating. *Corros. Sci.* **2015**, *90*, 239–247. [CrossRef]
33. Ramezanzadeh, B.; Haeri, Z.; Ramezanzadeh, M. A facile route of making silica nanoparticles-covered graphene oxide nanohybrids (SiO$_2$-GO); fabrication of SiO-GO/epoxy composite coating with superior barrier and corrosion protection performance. *Chem. Eng. J.* **2016**, *303*, 511–528. [CrossRef]
34. Liu, B.; Fang, Z.G.; Wang, H.B.; Wang, T. Effect of cross linking degree and adhesion force on the anti-corrosion performance of epoxy coatings under simulated deep sea environment. *Prog. Org. Coat.* **2013**, *76*, 1814–1818. [CrossRef]
35. Ramezanzadeh, B.; Ahmadi, A.; Mahdavian, M. Enhancement of the corrosion protection performance and cathodic delamination resistance of epoxy coating through treatment of steel substrate by a novel nanometric sol-gel based silane composite film filled with functionalized graphene oxide nanosheets. *Corros. Sci.* **2016**, *109*, 182–205. [CrossRef]
36. Westing, E.P.M.V.; Ferrari, G.M.; Wit, J.H.W.D. The determination of coating performance with impedance measurements—II. Water uptake of coatings. *Corros. Sci.* **1994**, *36*, 957–977. [CrossRef]
37. Behzadnasab, M.; Mirabedini, S.M.; Esfandeh, M. Corrosion protection of steel by epoxy nanocomposite coatings containing various combinations of clay and nanoparticulate zirconia. *Corros. Sci.* **2013**, *75*, 134–141. [CrossRef]
38. Ghazizadeh, A.; Haddadi, S.A.; Mahdavian, M. The effect of sol–gel surface modified silver nanoparticles on the protective properties of the epoxy coating. *RSC Adv.* **2016**, *6*, 18996–19006. [CrossRef]
39. Wang, G.; Zhu, L.; Liu, H.; Li, W. Zinc-graphite composite coating for anti-fouling application. *Mater. Lett.* **2011**, *65*, 3095–3097. [CrossRef]
40. Abdel-Aal, N.; Sawada, K. Inhibition of adhesion and precipitation of CaCO$_3$ by aminopolyphosphonate. *J. Cryst. Growth* **2003**, *256*, 188–200. [CrossRef]
41. Zhang, Z.; Chen, S.; Li, Y.; Li, S.; Wang, L. A study of the inhibition of iron corrosion by imidazole and its derivatives self-assembled films. *Corros. Sci.* **2009**, *51*, 291–300. [CrossRef]

© 2018 by the authors. Licensee MDPI, Basel, Switzerland. This article is an open access article distributed under the terms and conditions of the Creative Commons Attribution (CC BY) license (http://creativecommons.org/licenses/by/4.0/).

Article

IPN Polysiloxane-Epoxy Resin for High Temperature Coatings: Structure Effects on Layer Performance after 450 °C Treatment

Simone Giaveri [1,†], Paolo Gronchi [1,*] and Alessandro Barzoni [2]

1. Department of Chemistry, Material and Chemical Engineering, G. "Natta", Politecnico di Milano, P.za Leonardo da Vinci 32, 20133 Milano, Italy; simone.giaveri@mail.polimi.it
2. Akzo Nobel Coatings, Via Silvio Pellico, 10, 22100 Como, Italy; Alessandro.Barzoni@akzonobel.com
* Correspondence: paolo.gronchi@polimi.it; Tel.: +39-02-2399-3274
† Present address: Institute of Materials, École Polytechnique Fédérale de Lausanne, EPFL, Route Cantonale, 1015 Lausanne, Switzerland.

Academic Editors: Assunta Marrocchi and Maria Laura Santarelli
Received: 23 October 2017; Accepted: 22 November 2017; Published: 28 November 2017

Abstract: Coatings for high temperatures (HT > 400 °C) are obtained from interpenetrating polymer network (IPN) binders formed by simultaneous polymerization of silicone and epoxide pre-polymers. A ceramic layer; mainly composed of silica and fillers; remains on the metal surface after a thermal treatment at 450 °C. The layer adhesion and the inorganic filler's distribution have been investigated by, firstly, exchanging the organic substituents (methyl and phenyl) of the silicone chains and, secondly, by adding conductive graphene nanoplatelets with the aim to assure a uniform distribution of heat during the thermal treatment. The results are evidence that different substituent ratios affect the polymer initial layout. The adhesion tests of paint formulations are analysed and were related to instrumental analyses performed using glow discharge optical emission spectroscopy (GDOES); thermal analyses (TG/DTA and DSC); electron microscopy with energy dispersive X-ray analysis (SEM-EDX). A greater resistance to powdering using phenyl groups instead of methyl ones; and an improved distribution of fillers due to graphene nanoplatelet addition; is evidenced.

Keywords: high-temperature coatings; corrosion protection; powder coatings

1. Introduction

The protection against corrosion of a metal surface with an organic or inorganic coating is the main application for coatings [1] (the global anticorrosion coating market was estimated to be $24.84 billion in 2017 and is projected to reach $31.73 billion by 2022, at a CAGR (Compound Annual Growth Rate) of 5.0% from 2017 to 2022 [2]). An important demand in this sector is the protection against thermo-oxidative corrosion of the metal parts of the internal combustion engines, turbines, and heaters (other applications are: oven parts, chimney pipes, fireplace inserts, steam lines, furnaces, lighting fixtures, heat exchangers, boilers, engines, exhaust stacks, and mufflers). The wall temperature reaches 400–600 °C, at which point the organic binders degrade due to the chemical and thermal instability of the C–C bonds. Traditional organic coatings for metals, like those based on acrylic and/or epoxy polymers, are stable only up to 60–80 °C; above 150 °C, degradation takes place. On the contrary, the binders based on hybrid organic–inorganic-containing polysiloxane show superior thermal stability [3].

The thermal stability increases more if an interpenetrating polymer network (IPN), with the chains entangled one into the other in a dense network, is used as the binder [3]. Indeed, organic- and inorganic-based monomers, or pre-polymers, mixed together, can separately self-polymerize with different chemistries changing the material performances, an example is the thermal behaviour, from that obtained with the single resin one [4]. These binders, when they also contain ceramic or metallic fillers, find application for heat resistant coatings at high temperatures [4,5]. However, the binders previously described in the literature are intercrosslinked networks obtained with sequential reaction steps, not with a simultaneous polymerization, using networking agents and curing. In the first case, the compatibility between organic and hybrid polymeric materials is obtained by the formation of covalent bonds, but thermal stability appears lower than in the simultaneous IPN obtained by a single step.

The characterisation of silicone or IPN coatings, used as binders, relies on colour and gloss changes, blistering, cracking, and loss of adhesion after exposures at high temperature. Loss of gloss is produced by continuous break-down of the organic chains at the surface. However, when the phenomenon becomes so strong and extensive, until the complete degradation of the binder, the pigment splits from the binder and chalking occurs; chalking is the term used to define the release of pigments and particles [6,7].

The present work concerns the coatings obtained by a simultaneous polymerization of silicone and epoxy-acrylic pre-polymers to give IPNs [4], in which the formation of physical constraints, like polymeric domains or polymer-matrix nanocomposite distribution, affects the chalking extent. The constraints are supposed to affect the removal of thermally-degraded organic fractions from the coating surface. Silicone-epoxy-acrylic IPNs were selected according to their properties. Epoxy resins are characterized by low shrinkage, easy curing and processing, and the films obtained from them acquire excellent solvent and chemical resistance, great toughness, and good adhesive strength; otherwise, they suffer thermal stability and pigment holding ability [7]. Silicone resins and the derived films show superior thermal and thermo-oxidative resistance, partial ionic nature, excellent moisture resistance, low surface energy and good flame retardant properties. Moreover, silicone resins are used as epoxy resin modifiers [8]. Acrylics, properly reacting with epoxies, were added to improve the mechanical properties of the coatings. With the aim of increasing IPNs thermal resistance, we examined the properties of polymer-graphene nanocomposites to understand the behaviour of these materials at high temperatures. Indeed, following the indications of the literature [9–11], the high graphene thermal conductivity associated with its leaflet morphology, could promote a homogeneous thermo-oxidative degradation of the coatings [12]. Moreover, graphene's low permeability to all gases and salts, due to the quasi 2D structure, appears to be an excellent candidate in anticorrosion coatings [13].

Sample films were obtained by powder paints due to the high performances and environmental constraints, accompanied by a satisfactory and easy application [14–16]. Adhesion tests are performed and the results, examined also to infer the powdering degree, are interpreted by the data obtained with instrumental analyses: firstly, thermal analysis, and then GDOES and SEM.

2. Materials and Methods

2.1. Materials

The coating components were:

- As binder: (IPN polymer resin) the silicone resins (Wacker Silres®, Milano (MI), Italy) and Dow Corning Xiameter® (Midland, MI, USA) and epoxy resins (DOW D.E.R.®, Midland, MI, USA), were added with acrylic resins (BASF, Joncryl®, Ludwigshafen am Rhein, Rhineland-Palatinate, Germany);
- As fillers/pigments: baryte (Aprochimide Baryte, Muggiò (MI), Italy), wollastonite (Nyco Nyad, Hermosillo, Sonora, Mexico), micro mica (Norwegian Talc, N-5355 Knarrevik, Norway), manganese ferrite black spinel (Ferro, Cleveland, OH, USA), and graphene nanoplatelets (Directa Plus, Lomazzo, Italy);
- As additives: benzoin (Miwon Speciality Chemical, Gyeonggi-do, Korea), flow control additive (Estron Chemical, Calvert City, KY, USA) and fluidization additive (Evonik Industries, Essen, Germany).

2.2. Paint Formulations

Many different paint formulations were prepared (see [17] for a complete description of formulations and a listing of prepared samples). In this paper, we considered four of them using the compositions reported in Table 1.

The weight of each batch was 1 kg and the total ligand amount was 510 g.

Table 1. Raw materials (wt %) and sample composition of selected samples *.

Raw Materials	Coating Composition ID			
	S4	S6	SF1	SF2
Phenyl Silicone	25	–	–	–
Methyl-Phenyl Silicone	–	25	25	25
BPA Epoxy Resin	11	11	11	11
Carboxyl Acrylic Resin	15	15	15	15
Pigment Black	1.5	1.5	1	1.2
Fillers (1)	46.3	46.3	46.3	46.3
Additives (2)	1.2	1.2	1.2	1.2
Graphene nanoplatelets	–	–	0.5	0.3

*: Batch of 1 kg each: (1) micro mica, baryte, wollastonite; (2) degassing agent, flow control additive and fluidization additive.

The components of pigment and filler inorganic mixtures are presented in Table 2, together with the physical properties.

Table 2. Pigment and filler.

Pigments and Fillers	Chemical Analysis	Density (g·cm^{-3})	Oil Adsorption (g/100 g)	Particle Size (µm)
Pigment Black	MnFe$_2$O$_4$	4.5 [1]	48 [2]	0.5 [3]
Baryte	BaSO$_4$ (97%)–SiO$_2$ (2%)	4.35 [4]	11 [5]	4 [6]
Micro Mica	KAl$_2$(AlSi$_3$O$_{10}$)(OH)$_2$	0.5 [7]	49 [8]	6 [9]
Wollastonite	CaSiO$_3$	1.04 [10]	24 [11]	9 [3]

[1]: DIN-ISO 787 part/Teil 10; [2]: DIN-ISO 787 part/Teil 5; [3]: Median particle size; Cilas granulometer HR 850-B; [4]: ASTM D153-82; [5]: ASTM D281; [6]: Average particle size; [7]: Tamped density; ISO 787-11; [8]: ISO 787-5; [9]: Median particle size; [10]: ASTM C 87; [11]: ASTM C 87.

2.3. Paint Preparation

A detailed description of the powder coatings production process is outlined in the following list: Raw material weighting—Low temperature mixing—Extrusion—Calendering—Cooling down—Pelletizing—Milling—Sieving—Post-additivation—Sieving, again.

An OMC Saronno EBVP30/20 co-rotating intermeshing twin screw extruder (OMC, Saronno (MI) Italy) was used; right-handed and left-handed elements and kneading disks provided distributive and dispersive mixing, extremely important in the case of nanofillers [10]. During extrusion, the temperatures of the five screw regions were set at 15 °C, 30 °C, 90 °C, 110 °C, and 80 °C respectively; the feeding screw was set to 1.5 rpm, and torque was checked.

After calendering, −25 °C cooling was performed in order to simplify the pelletizing process. Polymeric pellets were milled by an ALPINE rod mill (HOSOKAWA ALPINE Aktiengesellschaft, Augsburg, Germany) and later sieved using an Endecotts laboratory test (London, UK), 125-µm sieve.

2.4. Application

Powder coatings were applied on standard low-carbon, cold-rolled steel (6 cm × 20 cm) panels after sandblasting with inert soda-lime and tempered glass microspheres, whose mean particle size was 214.6 µm. The treatment was performed manually adopting a gun-iron panel distance at about 30 cm.

Powder coatings application was performed manually using a corona charging system on a GEMA powder gun (GEMA, Trezzano sul Naviglio, Italy), according to [18]. Powder was carefully fed into a venturi pump and then pneumatically transported to the corona gun by a feeding gas at 1.3 bar pressure. One hundred kilovolts were applied at room temperature at a 250-mm gun tip-substrate distance. A flat nozzle was used, forming an elliptical powder cloud in front of the gun. Then, a 40–60 µm coating thicknesses were applied in order to maximize the thermal resistance. Curing was performed in a Thermo Scientific Heraeus batch oven (Thermo Fisher Scientific, Waltham, MA, USA) at 210 °C for 20 min. High-temperature tests (HT) were performed at 450 °C for 12 h.

2.5. Instrumental Methods

The thickness of the coated panels was measured by electromagnetic induction based on Byko test 8500 (Byko, Norvik hf, Reykjavik, Iceland). High-temperature samples with exposure at 450 °C for 12 h was performed. Paint formulations, coated on sandblasted panels, were placed in a Galli G-21HT muffle furnace (Galli, Fizzonasco di Pieve Emanele, Italy). After exposure, specimens were removed from the furnace to be air-cooled to room temperature for at least one hour. Inspection was carried out according to the ASTM D2485-91 (2013) standard. Evidence of peeling, cracking, blistering, abnormal discoloration, chalking, and loss of adhesion were checked. An adhesion cross-cut test was performed before and after heat treatment at 450 °C for 12 h according to ISO 2409:2007 (E). A six-blade cutting Byko-cut Universal Paint Inspection Gauge (Byko, Norvik hf, Reykjavik, Iceland) with 1 mm spaced cutting edges was used. Thermogravimetric and differential thermal analyses (TG–DTA) were performed with a SII Seiko Instruments Exstar 6000 TG–DTA 6300 unit (Seiko Instruments, Torrance, CA, USA). Thermal runs were set from 25 °C to 800 °C; an air atmosphere was used to study coatings' thermal degradation as in real working conditions. A heating rate of 10 °C·min^{-1} was selected. The results were reported showing wt % vs. temperature in thermogravimetric plots, as wt %·min^{-1} vs. temperature in DTG graphs, and as thermocouple µV vs. temperature in DTA curves.

Differential scanning calorimetry (DSC) analyses were performed with a heat flux SII Seiko Instruments Exstar 6000 (Seiko Instruments, Torrance, CA, USA). The selected thermal cycle was composed of a ramp from 25 °C to 300 °C at a 10 °C·min^{-1} heating rate, followed by cooling to 25 °C at the same speed. The heating was then repeated with a thermal run from 25 °C to 500 °C at 10 °C·min^{-1}; a nitrogen atmosphere was used. Results were reported as mW/°C. Finally, the areas under the peaks were calculated using machine software (Seiko Instruments Inc. "SII", Iwate, Japan).

Qualitative glow discharges optical emission spectroscopy (GDOES) was performed with a Spectruma GDA 750 analyser (Spectruma, Hof, Germany). Scanning electron microscopy-energy dispersive X-ray spectroscopy (SEM-EDS) analyses was performed using a Zeiss EVO 50 EP-Oxford Inca energy 200 LZ4 spectrometer (Carl Zeiss, Oberkochen, Germany).

3. Results and Discussion

Four IPN specimens were taken as representative among the many (the investigations were performed using more than 100 preparations [17] so as to cover many parameter effects); they were named as S4, S6, SF1, and SF2, respectively (Table 1). S4 and S6 are composed of a mixture of three resins (silicone, epoxy, and carboxylic acrylic) together to pigment black, functional fillers, and additives. A phenyl silicone resin is present in S4, while a methyl-phenyl silicone resin is present S6, with the same weight fraction. In SF1 and SF2 IPNs, the same S6 formulation was maintained, except for partial substitution of the pigment concentration by graphene nano-platelets, with different weight fractions. Functional fillers and additives were added to decrease the amount of pyrolytic decomposition of the organic binder and to avoid defect formation during the film-forming process, respectively [19].

3.1. Adhesion

In order to satisfactorily prevent substrate oxidation in thermal oxidative conditions, coatings must adhere to the surface on which they are applied [4,5]. Cross-cut tests were carried out to assess film adhesion before and after heat treatment.

The images of the coated surfaces printed on the tapes, removed from the cross-cut areas, are presented in Figures 1 and 2 in order to provide evidence of the flaking (powdering) degree. Optical microscopy has been used to obtain the high-definition images for a reliable evaluation. S6 and SF1 paint formulations show the best adhesion before heat treatment and their rating is consequently classified as 0 (Table 3 according to ISO 2409:2007 (E)); a different intensity of failures was found for the S4 and SF2 samples.

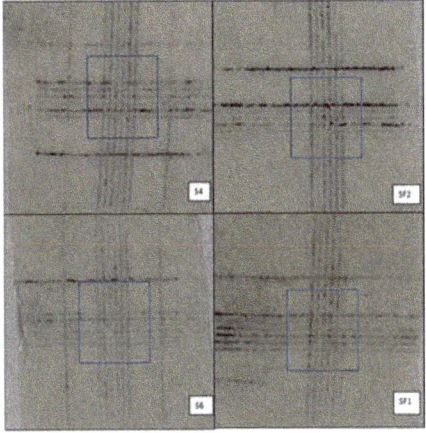

Figure 1. Prints of cross-cut area on tape before thermal treatment (the clearness is worsened by the glue layer).

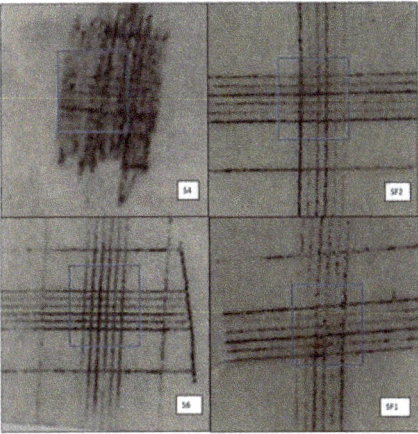

Figure 2. Prints of the cross-cut area on tape after the thermal treatment (the clearness is worsened by the glue layer).

Table 3. Sample classification on adhesion tests (according to ISO 2409:2007 E), before and after heat treatment (HT, 450 °C 12 h; air).

Sample	Classification		Assessment
	Before HT	After HT	
S4	1	2	Before HT: small flake separation (<5% of the analysed area); After HT: The coating has flaked along the edges and/or at the intersections of the cuts. Failure (5–15% of the analysed area)
S6	0	0	Before/after HT: the edges of the cuts are completely smooth; none of the squares of the lattice is detached
SF1	0	1	Before HT: the edges of the cuts are completely smooth; none of the squares of the lattice is detached; After HT: small flake separation (<5% of the analysed area)
SF2	2	1	Before HT: The coating has flaked along the edges and/or at the intersections of the cuts. Failure (5–15% of the analysed area). After HT: small flake separation (<5% of the analysed area).

After thermal treatment (the tape images, after heat treatment, are reported in Figure 2), a strong pyrolytic decomposition of the IPN organic part was expected. Indeed, although silicone resin backbone has high thermal stability, its organic substituents suffer strong degradation in air at $T > 300$ °C [20–24]. According to [20], the oxygen catalyses the silicone weight loss, the residual part of it (the backbone) undergoes the formation of a thermally-stable and tightly-reticulated, more inorganic than organic, network. Furthermore, the adhesion on the substrate may be promoted by the high temperature due to element inter-phase diffusion. Thus, cross-cut tests were carried out one more time to check the substrate coating adhesion and the silicone effect. The adhesion classification, before and after HT, is reported in Table 3.

The S4 samples have the weakest adhesion with a worsening after HT, while the S6 and SF1 coatings remain almost stable and the SF2 samples seems to show a small increase of the adhesion after heat treatment. Adhesion tests suggest that the phenyl amount on silicone chains have to be reduced: phenyl silicone S4 sample indeed shows a bad result on cross-cut test after heat treatment. In an attempt to explain the behaviour, we could hypothesize that carbon and carbonaceous residues would remain for a longer time in the silicon oxide network and interfere with the formation of bonds to the metal during pyrolysis due to slow combustion kinetic of aromatic substituent at 450 °C.

Indications of chalking can be inferred from the images of the adhesion tests; firstly, referring to the high degradation degree of the S4 sample (Figure 2), the tape surface does not present any transparency over the whole cutting area caused by the de-structured binder. In this case, the phenyl silicone resin shows a strong and intense chalking, so the cross-cut area is much darker in S4 tape in comparison with the same area of methyl-phenyl silicone-containing samples.

For the sake of completeness, in Figure 3 we report the images of S6 and SF1 panels after heat treatment. At the same adhesion, no shadows are present on the SF1 surface, both in and out of the cutting area, indicating a more limited chalking.

The presence of graphene probably increases adhesion after thermal treatment as the SF2 sample assessment is two steps before HT (Table 3, change from two to one). As expected, the chalking increases after heating as supported by the bad transparency of the tapes, as reported in Figure 2. However, from the visual exam of the panel instead of the tapes after HT (Figure 3), SF1 presents fewer shadows than S6 indicating a less chalking degree. Investigating the filler amount effect, SF1 and SF2 present little and no discernible cleanliness difference on tapes.

The superior resistance to chalking produced by graphene addition may be related to enhanced properties of polymer-graphene nanocomposites [10]. The network of graphene nanoplatelets inside the binder matrix is supposed to increase mechanical and thermal performances at high temperatures [9,11]. Finally, thermal conductivity of the graphene sheet may favor the homogeneous degradation of the organic part of the IPNs binder, avoiding high-temperature peaks [4].

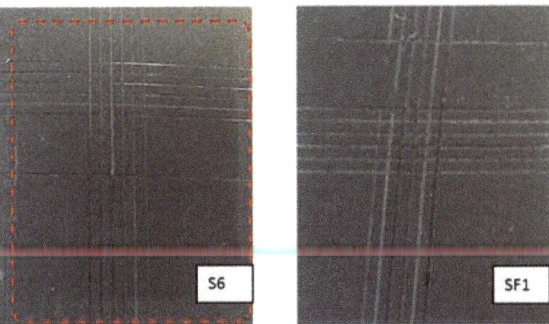

Figure 3. Comparison between S6 vs. SF1 coated panels after thermal treatment.

3.2. Thermal Analysis

The thermo-gravimetric analysis (TG) provides information on thermal stability (weight loss at increasing temperature) of coatings after heat curing (210 °C) and suggests hypotheses about the degradation mechanism. All curves are characterized by two weight losses (Figure 4) according to the cited literature with different rates of degradation.

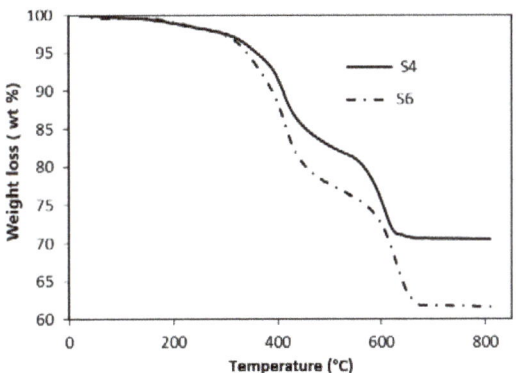

Figure 4. Comparison of the curves describing the normalized weight losses of the S4 and S6 samples.

The first and the second weight loss (wt %) are reported in Table 4 relative to the four examined samples: the total losses are in the 26–37.3% range at high temperatures (680 °C, max). The result is related not only to the presence of a high amount of functional fillers in each formulation, as reported in Table 1, but also to the presence of silicone polymers that have high residual masses after thermal degradation (up to 55% for methyl-phenyl [24]). Nevertheless, the oxidative cross-linking reaction rearranges all the residual organosilanes to silica. The reaction mechanism is probably triggered by the formation of radicals on side groups that react with oxygen and produce peroxide functions which accelerate the rearrangement. Finally, in an oxidative environment, some silica may be obtained from the competition between volatilization of oligomers and oxidative cross-linking [20–22,25].

Table 4. Coating weight losses (%).

Steps	T (°C)	Weight Loss (wt %)			
		S4	S6	SF1	SF2
First step	280–530	15.7	21.6	22.7	20.6
Second step	530–680	11.1	14.8	14.6	12.4
Total		26.8	36.4	37.3	33.0

The S4 sample shows the lowest total wt %, mainly concentrated at low temperatures. S6 and SF1 are characterized by higher weight losses than S4, while SF2 presents an intermediate total wt %. The similarity of the second step wt % between S6 and SF1 allowed us to conclude that weight loss does not depend on graphene. We have to note that, from the thermogravimetric point of view, the SF2 losses are less than both S6 and SF1, and are probably due to a different formulation (more pigment is present, as reported in Table 1).

The presence of two strong weight losses were highlighted by time differentiating the TG curves of S4 and S6 samples, as reported in Figure 5.

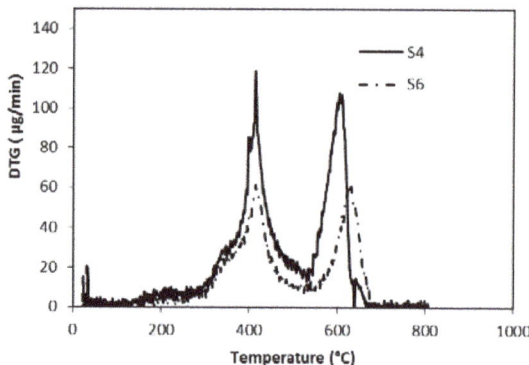

Figure 5. Comparison of the curves describing the degradation rates of the S4 and S6 samples (5.0 mg and 2.5 mg samples, respectively; not normalized by weight).

For both of these samples, two maxima appear at 413 °C and 630 °C, respectively. The first ones have a similar shape (with a shoulder before the peak) and the same maximum temperatures at 413 °C, suggesting the same thermal effect characterized by a beginning lower than 300 °C that could be attributed to the depolymerisation and bond breakings of epoxy-acrylic organic binder fraction. Therefore, DTG analysis revealed that thermal behaviour of epoxy-acrylic binder is similar, notwithstanding the different substituents (phenyl, instead of methyl groups). The only difference is the shoulder that has a different intensity is probably due to the phenyl group fraction.

Once again, the phenyl group seems responsible for the significant temperature differences attributed to the oxidation of organic substituents of silicones, highlighted for the second peak: this appears 27 °C before in the S4 formulation than in S6. Thermal analysis is in agreement with the literature for PDMS stability [20–24], even when considering the presence of other components inside formulations. The temperature difference between the two high-temperature DTG peaks may be attributed to different organic parts, as S4 silicon resin has phenyl groups only, while S6 has methyl and phenyl substituents. The introduction of phenyl substituents on silicone chains has been found to increase the onset temperature of degradation [21,22]. According to [23], the thermal stability of siloxane polymers increases on decreasing the phenyl content and this trend is confirmed by our analyses (the maximum rate of the weight loss on S4 sample appears at ca. 580 °C instead of ca. 600 °C for the S6 sample). Analysing the thermal behaviour at higher temperatures, during polydiphenyl-dimethyl siloxane pyrolysis, evolution of free benzene and volatile cyclic siloxanes occurs, as reported in [20,22]. As seen in Figure 4, the carbon removal, during silicone thermal degradation appears more pronounced at the 280–530 °C weight loss because of a decrease in the phenyl content (a difference of 5.9%); this effect may be due both to the different steric hindrances and to the layout of silicone chains due to methyl of phenyl groups [23]. Accordingly, the inspections of residues of polydiphenyl-dimethyl siloxanes thermally degraded in air revealed a greyish white powder constituted mainly of white silica and black silicon-oxycarbide [23,25] in which Si atoms are simultaneously bonded to carbon and oxygen [26]. Moreover, from the elemental analysis, the weight percent of C concentrations inside residues has been found to be three times higher in the case of a phenyl polysiloxane with respect to a fully-methylated one [23]. Hence, as the phenyl content decreases, a more stable and protective Si–O–C layer is formed without aromatic ring interferences, with low carbon content [24]. As a conclusion, this hypothesis is in accordance with thermogravimetric analysis, which clearly indicates that the S6, SF1, and SF2 samples, composed by a methyl-phenyl silicone, show a greater weight loss than S4. Specifically, thermal analyses indicate that S4 has a lower weight loss than S6, in agreement with the difficulty of carbon residues' oxidation/separation, as hypothesized above.

Finally, the S4 residue was supposed to present high C content and low ceramic yield. This hypothesis is strongly supported by the lower adhesion and stronger chalking exhibited by the S4 coating with respect to S6, as discussed above [23,24]. Moreover, the second DTG peak is about 30 °C before for the phenyl substituted silicon chain, probably meaning that the oxidized fragments are placed in the upper part of the film, in a more favourable position for chalking to occur.

S6, SF1, and SF2 formulations are compared in Figure 6; the shape of the curves is the same for the three coatings. From the DTG plot, reported in Figure 7, it is evident that the amount of graphene is not relevant as the weight losses occur at approximately the same temperatures. In comparison with graphene containing formulations, the shape of the DTG curves of S6 are almost the same as SF1 and SF2, with the exception that the rates of weight losses of SF1 and SF2 (these curves are superimposed in Figure 7), graphene-containing formulations, occur 15 °C before those of the S6 sample, probably due to a different material thermal conductivity.

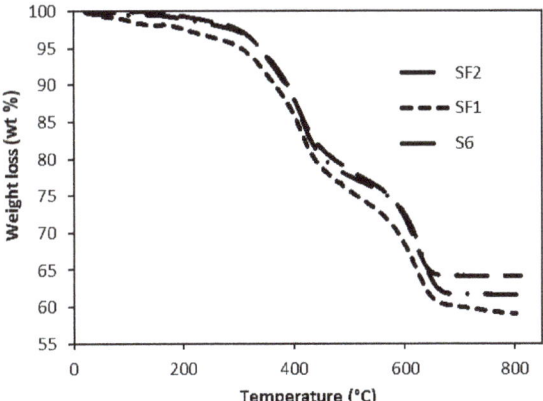

Figure 6. Comparison of the weight losses of the S6 sample, without graphene, and the SF1 and SF2 samples, with graphene.

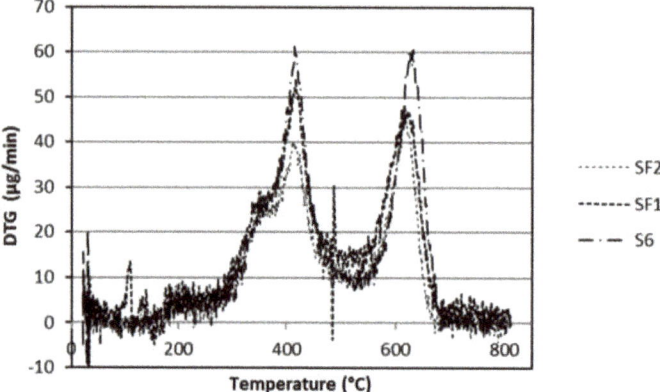

Figure 7. DTG curves of S6, SF1, and SF2 samples (not normalized).

Differential scanning calorimetry (DSC) analyses for the same samples, performed until 300 °C, are reported in Figure 8. The curves are placed at different µW level and have a similar outline, probably related to the graphene amount. In fact, SF1 and SF2 curves appear flatter mainly after 200 °C, meaning a stability of the specific heat, probably related to graphene amount because of the slope of the curve, decreases in the order: S6 (0% graphene) > SF2 (0.3% graphene) > SF1 (0.5% graphene).

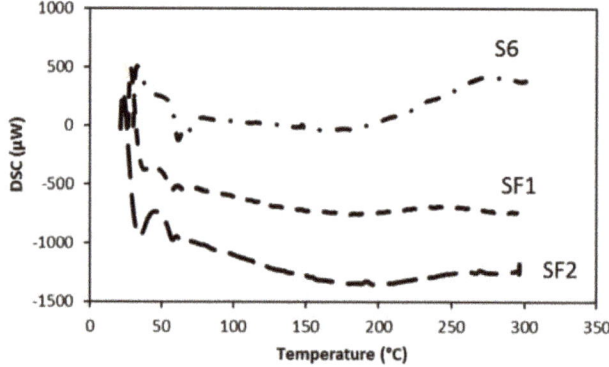

Figure 8. Comparison of the DSC behaviors of S6, SF1, and SF2 samples at the same weight (5 mg).

3.3. Film Chemical Composition

We analysed the S4 and S6 coatings with the Glow discharge optical emission spectroscopy (GDOES). The difference between the two samples is the organic part of silicone binder: methyl (S4) and phenyl-methyl (S6), respectively (Table 1). The analysis allows checking if some layering had taken place due to the inhomogeneous distribution of pigments and fillers all along the coating thickness. Results are plotted in Figures 9 and 10.

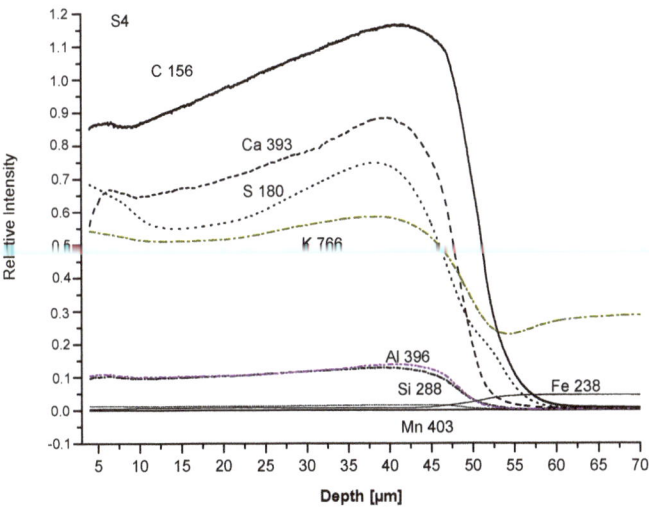

Figure 9. GDOES spectra of the S4 sample.

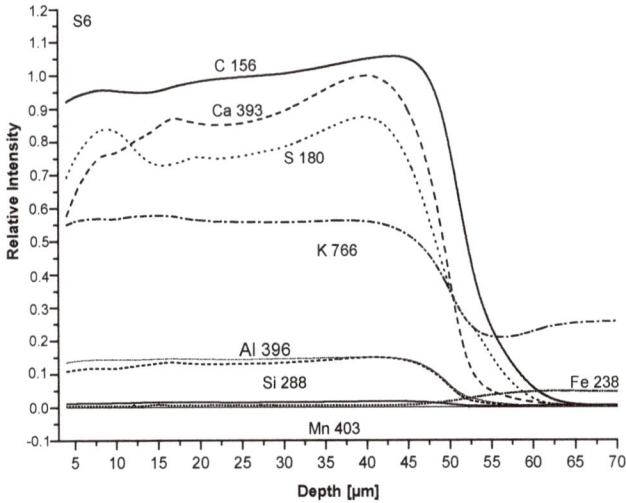

Figure 10. GDOES of the S6 sample.

Firstly, the film thickness appears properly produced, between 40 and 60 μm. Moreover, no relevant layering is observed in both coatings as the concentration of all elements from the film surface to the substrate both slightly change at the same ratio (C, Ca, S) or remain unchanged (K, Al, Si, Mn). The first group of elements is characterized by concentration gradients; C (from the binder) and Ca appear more densely packed near the substrate, particularly in the S4 than in the S6 sample. Otherwise, the S line, related to the baryte concentration, appears more elevated both at the surface and at the bottom of the film than in the middle. Inversely, S4 and S6 analyses show that the curves related to Mn are nearly horizontal, confirming the homogeneous dispersion of pigment inside the coatings. Similarly, K, Si, and Al concentrations, related to micro mica, appear quite stable. To conclude, the comparison between S4 and

S6 coatings indicates a greater homogeneity and lack of important gradients of concentration, this is true of S4 in the S6 composition. For chalking, the GDOES analysis seems to confirm that it depends on the release of pigment particles or extenders (particularly baryte) after the thermal degradation of the binder; no relevant layering of pigment and/or filler appears inside the coating to justify the difference in the chalking performance. From the other site, the increasing binder concentration from the surface towards the substrate, which is very pronounced in the S4 sample, could have caused a deeper degradation and an increase of the particle amount released at a constant concentration of all the other elements.

3.4. SEM Analysis

Coating morphology was analysed by scanning electron microscope (SEM) and energy dispersive spectroscopy (EDS). S4 and S6 SEM micrographs, before HT, are reported in Figures 11 and 12.

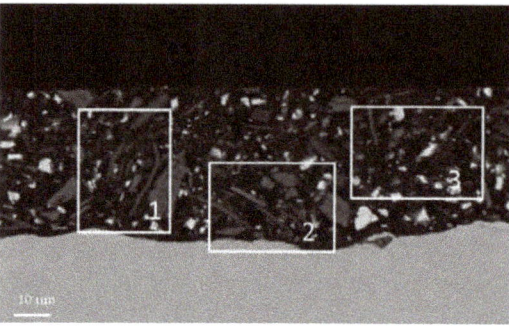

Figure 11. S4 coating; section before HT. In the squares: (1,2) densely packed wollastonite and (3) wollastonite free region.

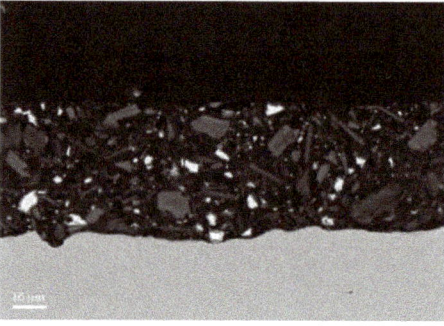

Figure 12. S6 coating; section before HT. Homogeneous distribution of fillers.

The SEM micrographs of the coating section before the heat treatment help to analyse the distribution of fillers in the film. White spots were found to be baryte, large light grey ones', wollastonite, and dark grey ones micro mica. By comparing the distribution of the large wollastonite particles it appears that the filler distribution is more homogeneous in the S6 section (Figure 12). All the particles, however, are (a) more homogeneously distributed in the S6 film than in the S4 one supporting the chalking results; and (b) denser and not homogeneously packed in S4 sample near the substrate surface than in the S6 one, validating the GDOES results.

Polymeric domains, with a resembling sphere geometry, were identified in the S6 and not in the S4 coating. Images are reported in Figure 13. From EDS analysis, the domains are rich of silicone inside IPNs. According to [27], the uniformly dispersed dark spherical agglomerates might be formed by condensed polyhedral oligomeric 3D siloxane structures. These may be promoted by the presence of low steric hindrance methyl groups, allowing rearrangements of macromolecules. These domains are reinforcing agents and they can properly affect properties of organic-inorganic hybrid composites. Hence, by heat treatment, agglomerates may be converted into thermally-stable silica particles, enhancing substrate adhesion and film cohesiveness. Additionally, the complexity of paint formulations prevents a more precise understanding of the phenomena. In fact, as reported in [23], the presence of acidic or basic impurities of fillers and residual catalysts might influence the polysiloxane thermo-oxidative degradation mechanism. The choice of the most performant coatings was, however, aimed to link the chalking effect to the film surface morphology after thermal degradation.

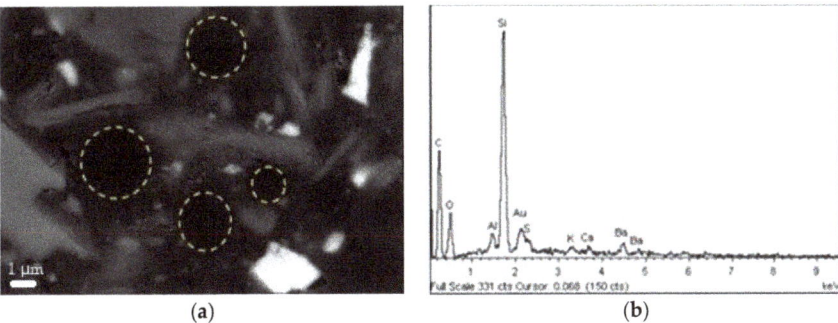

Figure 13. (a) SEM-EDS images at high magnification of S6 coating before HT. SEM; and (b) EDS analysis of S6 details, highlighted by yellow circles in the SEM image.

After 12 h at 450 °C, we examined S6 and SF1 samples by SEM-EDS; the low-magnification images are reported in Figure 14.

The two images at low magnification give a large sampling area and they clearly show large coating failures. It is evident that SF1 has a superior resistance to chalking in comparison with the S6 sample as the integrity of the coating is preserved in the SF1 sample, but not in the S6 one. Examining more deeply the images of Figure 14, the SF1 coating presents no delamination, blistering, and void formation, while the S6 sample is broken in several points of the coating.

At the end of our investigation, SEM images clearly indicate a greater resistance to deformation and breaking of the graphene containing coatings with respect to the S-type formulations. Superior properties of polymer-graphene nanocomposites are reported in literature [9–11,13]. Moreover, during thermal degradation of polysiloxanes, evolution of free benzene and volatile cyclic oligomers occurs [19,20,22].

According to [22], the importance of mass transfer of degradation products away from the coating is significant and it controls the degradation process. Hence, being that graphene has low permeability to all gases [13], it may promote the formation of a nano-structured ceramic residue, upon thermal conversion [26], leading to higher surface integrity, as revealed in Figure 14.

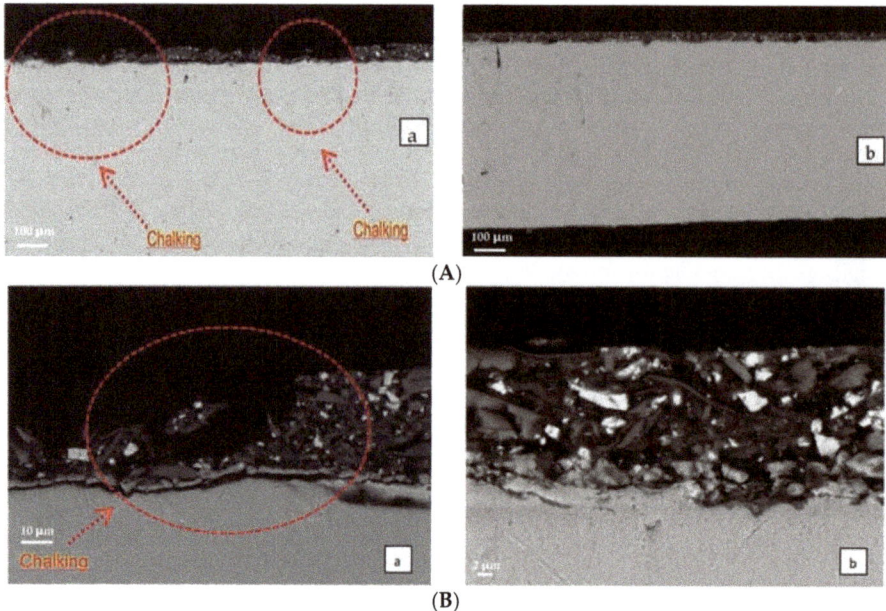

Figure 14. (**A**) Low- and (**B**) high-magnification images of sections after HT: (**a**) S6 coating section; (**b**) SF1 coating section.

4. Conclusions

In this paper we investigated the surface powdering after high-temperature treatment of IPN silicon-epoxide coatings. At first, the data led us to suppose that the IPN binder that is formed by two homopolymers braided to form a mesh causes the polymer chains to be well organized, without forming high-density domains and without forming intricate arrangements, like long chains typically shaping themselves. Indeed, our results provide evidence that, after the treatment at 450 °C, the organization of the inorganic part depends on the initial layout of the polymer chain. A greater homogeneity develops when the aromatic (phenyl) substituents of the silicone are substituted by methyl groups. Moreover, to obtain a deeper understanding, graphene was added because of its thermal conductivity, which more homogeneously distributes the temperature gradients inside the coating.

The adhesion test analyses reveal that the methyl groups promote a better adhesion of the paint residues on the metal substrate, after the treatment at 450 °C/10 min of the coatings, than the phenyl group. In a parallel manner, the powdering degree is reduced. Thermal analyses support the previous observations and reveal that the carbon residues of the phenyl group remain longer in the layer, with the detrimental effect. Adhesion and thermal analyses mainly refer to the macro-level observations, while both GDOES and the electron microscope analyses substantiate at a micro-level that a greater homogeneous distribution of fillers is obtained with graphene. The distribution of the residue after the high-temperature treatment appears a consequence of the polymer chain layout soon after the application. Parameters that need to be investigated further are as follows: the different organisation induced on the silicone resins by aromatic substituents in comparison to the methyl ones, the different organisation of the chain arrangement physically induced by graphene platelets, and, moreover, the different organisation induced by a better heat distribution during the high-temperature treatment. The likelihood of driving the coatings for high thermo-oxidative protection toward better performance seems to depend on the chemical and physical control of the thermal decomposition of the organic parts.

Acknowledgments: We acknowledge Akzo Nobel Coatings (S.P.A.), Como, CO, Italy, for the technical support and the materials supply.

Author Contributions: Simone Giaveri performed the experiments. Paolo Gronchi contributed analysis tools, analyzed the data, and wrote the paper. Alessandro Barzoni conceived and designed the experiments.

Conflicts of Interest: The authors declare no conflicts of interest.

References

1. Bohm, S. Graphene against corrosion. *Nat. Nanotechnol.* **2014**, *9*, 741–742. [CrossRef] [PubMed]
2. Biller, K. Anticorrosion coating market to reach $31.73 bn by 2022. *Focus Powder Coat.* **2017**, *2017*, 6.
3. Dhoke, S.K.; Palraj, S.; Maruthan, K.; Selvaraj, M. Preparation and characterization of heat-resistant interpenetrating polymer network (IPN). *Prog. Org. Coat.* **2007**, *59*, 21–27. [CrossRef]
4. Sperling, L.H.; Hu, R. Interpenetrating polymer networks. In *Polymer Blends Handbook*; Utacki, L.A., Wilkie, C., Eds.; Springer: Berlin, Germany, 2014; pp. 677–724.
5. Kumar, S.A.; Narayanan, T.S.N.S. Thermal properties of siliconized epoxy interpenetrating coatings. *Prog. Org. Coat.* **2002**, *45*, 323–330. [CrossRef]
6. Dhoke, S.K.; Maruthan, K.; Palraj, S.; Selvaraj, M. Performance of black pigments incorporated in interpenetrating polymer network (IPN). *Prog. Org. Coat.* **2006**, *56*, 53–58. [CrossRef]
7. Buxbaum, G. Introduction. In *Industrial Inorganic Pigments*, 2nd ed.; Wiley–VCH Verlag GmbH: Weinheim, Germany, 1998.
8. Ahmad, S.; Gupta, A.P.; Sharmin, E.; Alam, M.; Pandey, S.K. Synthesis, characterization and development of high performance siloxane-modified epoxy paints. *Prog. Org. Coat.* **2005**, *54*, 248–255. [CrossRef]
9. Kuilla, T.; Bhadra, S.; Yao, D.; Kim, N.H.; Bose, S.; Lee, J.H. Recent advances in graphene based polymer composites. *Prog. Polym. Sci.* **2010**, *35*, 1350–1375. [CrossRef]
10. Naebe, M.; Wang, J.; Amini, A.; Khayyam, H.; Hameed, N.; Li, L.H.; Chen, Y.; Fox, B. Mechanical property and structure of covalent functionalised graphene/epoxy nanocomposites. *Sci. Rep.* **2014**, *4*, 4375. [CrossRef] [PubMed]
11. Tong, Y.; Bohm, S.; Song, M. Graphene based materials and their composites as coatings. *Austin J. Nanomed. Nanotechnol.* **2013**, *1*, 1003.
12. Teng, C.; Ma, C.M.; Lu, C.; Yang, S.; Lee, S.; Hsiao, M.; Yen, M.; Chiou, K.; Lee, T. Thermal conductivity and structure of non-covalent functionalized graphene/epoxy composites. *Carbon* **2011**, *49*, 5107–5116. [CrossRef]
13. Potts, J.R.; Dreyer, D.R.; Bielawski, C.W.; Ruoff, R.S. Graphene-based polymer nanocomposites. *Polymer* **2011**, *52*, 5–25. [CrossRef]
14. Barletta, M.; Lusvarghi, L.; Pighetti Mantini, F.; Rubino, G. Epoxy-based thermosetting powder coatings: Surface appearance, scratch adhesion and wear resistance. *Surf. Coat. Technol.* **2007**, *201*, 7479–7504. [CrossRef]
15. Belder, E.G.; Rutten, H.J.J.; Perera, D.Y. Cure characterization of powder coatings. *Prog. Org. Coat.* **2001**, *42*, 142–149. [CrossRef]
16. Lee, S.S.; Han, H.Z.Y.; Hilborn, J.G.; Manson, J.E. Surface structure build-up in thermosetting powder coatings during curing. *Prog. Org. Coat.* **1999**, *36*, 79–88. [CrossRef]
17. Giaveri, S. High Temperature Organic Powder Coatings: Characterisation and Innovations. Master's Thesis, Politecnico di Milano, Milan, Italy, September 2015.
18. Meng, X.; Zhang, H.; Zhu, J. Characterization of particle size evolution of the deposited layer during electrostatic powder coating processes. *Powder Technol.* **2009**, *195*, 264–270. [CrossRef]
19. Grundke, K.; Michel, S.; Osterhold, M. Surface tension studies of additives in acrylic resin-based powder coatings using the wilhelmy balance technique. *Prog. Org. Coat.* **2000**, *39*, 101–106. [CrossRef]
20. Camino, G.; Lomakin, S.; Lazzari, M. Polydimethylsiloxane thermal degradation Part 1. Kinetic aspects. *Polymer* **2001**, *42*, 2395–2402. [CrossRef]

21. Mazhar, M.; Zulfiqar, M.; Piracha, A.; Ali, S.; Ahmed, A. Comparative thermal stability of homopolysiloxanes and copolysiloxanes of dimethyl/diphenyl silanes. *J. Chem. Soc. Pak.* **1990**, *12*, 225–229.
22. Deshpande, G.; Rezac, M.E. The effect of phenyl content on the degradation of poly (dimethyl diphenyl) siloxane copolymers. *Polym. Degrad. Stab.* **2001**, *74*, 363–370. [CrossRef]
23. Deshpande, G.; Rezac, M.E. Kinetic aspects of the thermal degradation of poly (dimethyl siloxane) and poly (dimethyl diphenyl siloxane). *Polym. Degrad. Stab.* **2002**, *76*, 17–24. [CrossRef]
24. Zhou, W.; Yang, H.; Guo, X.; Lu, J. Thermal degradation behaviours of some branched and linear polysiloxanes. *Polym. Degrad. Stab.* **2006**, *91*, 1471–1475. [CrossRef]
25. Narisawa, M. Silicone resin applications for ceramic precursors and composites. *Materials* **2010**, *3*, 3518–3536. [CrossRef]
26. Bernardo, E.; Fiocco, L.; Parcianello, G.; Storti, E.; Colombo, P. Advanced ceramics from preceramic polymers modified at the nano-scale: A review. *Materials* **2014**, *7*, 1927–1956. [CrossRef] [PubMed]
27. Pantano, C.G.; Singh, A.K.; Zhang, H. Silicon oxycarbide glasses. *J. Sol-Gel Sci. Technol.* **1999**, *14*, 7–25. [CrossRef]

© 2017 by the authors. Licensee MDPI, Basel, Switzerland. This article is an open access article distributed under the terms and conditions of the Creative Commons Attribution (CC BY) license (http://creativecommons.org/licenses/by/4.0/).

Article

The Potential of Functionalized Ceramic Particles in Coatings for Improved Scratch Resistance

Caterina Lesaint Rusu [1], Malin Brodin [2], Tor Inge Hausvik [3], Leif Kåre Hindersland [3], Gary Chinga-Carrasco [2], Mari-Ann Einarsrud [1] and Hilde Lea Lein [1,*]

[1] Department of Materials Science and Engineering, NTNU, N-7491 Trondheim, Norway; caterina.l.rusu@ntnu.no (C.L.R.); mari-ann.einarsrud@ntnu.no (M.-A.E.)
[2] RISE PFI AS, 7034 Trondheim, Norway; malin.brodin@rise-pfi.no (M.B.); gary.chinga.carrasco@rise-pfi.no (G.C.-C.)
[3] Berry Alloc, 4580 Lyngdal, Norway; TorInge.Hausvik@berryalloc.com (T.I.H.); LeifKare.Hindersland@berryalloc.com (L.K.H.)
* Correspondence: hilde.lea.lein@ntnu.no; Tel.: +47-73-55-08-80

Received: 28 May 2018; Accepted: 16 June 2018; Published: 19 June 2018

Abstract: The top layer of a typical high pressure floor laminate (HPL) consists of a melamine formaldehyde (MF) impregnated special wear layer (overlay) with alumina particles. This top layer plays a crucial role in determining the mechanical properties of the laminate. For HPLs, scratch resistance and scratch visibility are particularly important properties. This study aimed to improve the mechanical properties, particularly the scratch resistance, by adjusting the composition of the overlay. Laminates containing alumina particles were prepared and tested. These alumina particles were additionally functionalized with a silane coupling agent to ensure better adhesion between the particles and the resin. The functionalized particles led to enhanced scratch resistance of the laminates as well as improved dispersion of the particles within the resin. Micro scratch testing revealed that by using functionalized particles, the scratch surface damage was reduced and the recovery characteristics of the surface layer were improved. Higher scratch resistance and scratch hardness were thus obtained, along with a reduced scratch visibility.

Keywords: high pressure laminates (HPL); overlay; alumina; functionalization; silane coupling agent; scratch resistance; scratch visibility; scratch hardness

---

1. Introduction

Decorative laminates are widely used products with many applications within both home interior design and commercial settings. High pressure laminates are the most commonly used laminates consisting of multiple layers of kraft papers impregnated with phenol formaldehyde (PF) resin, laminated together with decorative paper impregnated with melamine formaldehyde (MF) [1]. For decorative laminates, the top surface has a double role: providing the final product with an excellent appearance and protecting the substrate from external damage [2]. Even a few scratches, though most of them are a few micrometers in depth, can deteriorate the original appearance [3] and in some cases, may even affect functionality by rendering them more vulnerable to other types of strain (flexion, impact, fatigue, etc.) [4]. Hence, scratch resistance and scratch visibility are very important factors for the top functional coatings in laminate floors. The melamine impregnated high-quality overlay paper based on α-cellulose [5] used on top of the decorative layer contains alumina particles to provide high scratch and abrasion resistance [1] to the finished product.

For coating applications such as laminate flooring where the appearance is particularly important, scratch visibility is a very significant issue [6]. Scratch visibility is essentially caused by the scattering of incident light due to high roughness or an uneven surface structure prompted by scratching (e.g., cracks, crazes, etc.) [7]. When these flaws on the damaged surface are larger than the wavelength of visible light, the scratches become visible [7]. The scratch visibility is known to correlate well with parameters like scratch hardness, scratch depth, surface roughness on the scratch path [6], coefficient of friction of the surface, and filler–matrix interface strength as well as the elastic recovery characteristics of the polymer [4]. The damage caused by plastic flow generally has a lower impact on the scratch visibility than brittle damage due to less scattered light [8,9].

In general, the addition of ceramic particles into a polymeric matrix is beneficial for scratch resistance [10,11] and depends on the type, morphology, and size of the particles [12–16], but also on their amount, hardness, and dispersion state [17]. Alumina has successfully been used as reinforcement for organic coatings with improved scratch resistance in a number of previous studies, for example, when added to Nylon-11 [15], melamine films [18] or PTFE [19]. A further improvement in the scratch resistance can be obtained through the strengthening of the polymer/particle interface [4], which will improve the dispersion state of the particles within the polymer matrix [20] and reduce the stresses associated with polymerization shrinkage [21] among others. Silane coupling agents are commonly used for functionalization, along with amines, carboxylates, phosphonates, etc., [22] as they can act as chemical bridges between the particle and the polymer. The silane can be covalently bound to the particle's surface through the silicon atom, and the alkyl chain will bond to the polymer. Important improvements in scratch resistance while maintaining good transparency have been achieved in the presence of ceramic nanoparticles such as silica [20], titania [23], and alumina [24] functionalized with silane coupling agents. Silica nanoparticles functionalized with three different trialkoxysilane agents and incorporated in polyacrylate yielded nanocomposite films with higher scratch and wear resistance as well as improved viscoelastic properties [20]. Titania nanoparticles functionalized with 3-aminopropyltrimethoxysilane (APTMS) greatly improved the resistance to abrasive wear and the particle dispersion within a Polyamide 11 matrix [23]. Better scratch resistance along with a good dispersion of the particles has also been obtained when incorporating alumina nanoparticles functionalized with [2-(3,4-epoxycyclohexyl)ethyl]trimethoxysilane in a polyurethane coating [24]. A significant improvement of the critical load, a uniform dispersion of the particles in the polymer as well as low residual scratch depths when compared to the neat polymer were obtained for all functionalized alumina containing samples even at relatively low particle loadings (6 vol.%) [24]. As such, a general trend in enhanced mechanical properties like wear resistance and resistance to scratch has been observed when incorporating functionalized nanoparticles into a polymer matrix.

For micron scale ceramic particles, the influence of functionalization on the properties of polymer composites has yet to be extensively studied. Additionally, there is a need for further work on the use of functionalized ceramic particles in laminate flooring in general and functionalized alumina particles in particular. The aim of this work was therefore to improve the mechanical properties of floor laminates, and especially the resistance to scratch through the addition of functionalized ceramic particles. Alumina particles were chosen as these particles are included in the overlay of commercial laminates. Today's laminates contain rather large alumina particles (~100 µm); however, these are hard to functionalize due to the size and the low surface area, and smaller particles are therefore preferred. Sub-micron particles are chosen over nano sized particles due to their advantages in terms of cost and availability. A silane coupling agent was selected for the functionalization of the alumina particles based on its compatibility with both the alumina particles and the MF polymer matrix. An improvement in the dispersion of the alumina particles as well as in the scratch resistance of the laminates were obtained through the addition of functionalized alumina particles, and the influence of the functionalization of alumina particles on the scratch resistance of laminates is discussed by evaluating different aspects such as residual scratch depth and width, scratch hardness, scratch visibility, and scratch deformation mode.

2. Materials and Methods

2.1. Materials

A commercial grade of MF resin (Dynea, Lillestrøm, Norway) in the form of a pre-condensate solution with a viscosity of 24 mPa·s at 25 °C, a dry content of 50 wt %, and pH 9.5 was used for the impregnation of the overlay paper. Overlay paper, decorative paper impregnated with MF and kraft papers impregnated with PF resin (supplied by Nordic Paper, Greåker, Norway) were used for manufacturing the laminates. A slip agent (formula and manufacturer unknown), wetting agent (Poly(oxy-1,2-ethanediyl),alfa-(2-propylheptyl)-omega-hydroxyl- INEOS Melamines GmbH, Frankfurt am Main, Germany), and curing agent (formula and manufacturer unknown) were used as additives for the preparation of the overlay impregnation solution.

Alumina particles (α-aluminum oxide, Al_2O_3, purity 99.99%) with an average size ranging between 0.5 and 1 μm (Inframat Advanced Materials, Manchester, CT, USA) were used as reinforcement for the MF resin. The average size was provided by the manufacturer. The specific surface area of the alumina powder was measured by nitrogen adsorption analysis to be 6.3 ± 0.12 m^2/g. The morphology of the particles obtained through SEM is shown in Figure 1a. The particles were irregularly shaped and their sizes corresponded to the range reported by the supplier.

(γ)-Glycidoxypropyl-trimethoxy silane (>98.0%), also referred to as AP-Silane 51, (Advanced Polymer, Inc., Carlstadt, NJ, USA) was used for surface functionalization of the alumina particles. The compound was selected based on its compatibility with both the alumina and the MF resin. The chemical structure of the compound is shown in Figure 1b. Tetrahydrofuran (THF, anhydrous, containing 250 ppm butylated hydroxytoluene (BHT), purity ≥99.9%, Advanced Polymer, Inc., Carlstadt, NJ, USA) was used as the solvent for the functionalization of the particles. BHT (200–300 ppm) was added to THF by the manufacturer to prevent the formation of explosive organic peroxides during storage [25].

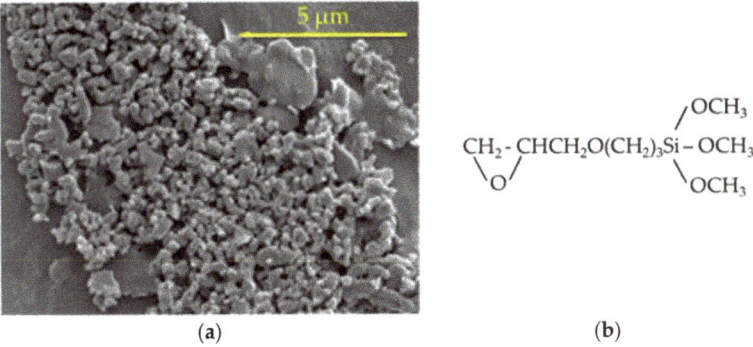

Figure 1. (a) SEM image of the alumina particles; (b) Chemical structure of the silane coupling agent used in this work, (γ)-Glycidoxypropyl-trimethoxy silane [26].

2.2. Methods

Surface functionalization of the alumina particles was carried out using the following procedure reported elsewhere [27] with slight modifications: alumina particles (8 g) were added to THF (25 mL) and the mixture was ultrasonically stirred for 15 min using a Bandelin Sonorex RK 255 Transistor bath operating at 35 Hz (Bandelin, Berlin, Germany). The dispersion was then added to the treatment solution containing the silane agent (4 g) dissolved in THF (25 mL), after which the resulting colloidal dispersion was stirred ultrasonically for 1 h at room temperature and subsequently separated by sedimentation. Finally, the sedimented particles were rinsed three times with THF to remove the organic residuals, then dried in a vacuum oven at room temperature [27]. Thermogravimetric analysis

(TGA) was performed on the as-received and functionalized alumina particles using a Netzsch thermal analysis system 4 (STA449, Erich NETZSCH GmbH & Co. Holding KG, Selb, Germany). The temperature range was 30–700 °C with a heating rate of 10 °C/min. Samples of approximately 30 mg were weighed and added to the alumina crucibles. Synthetic air was used as the flow medium with a flow rate of 25 mL/min.

Five types of laminates were manufactured by assembling a MF impregnated overlay paper containing alumina particles with a MF pre-impregnated decorative paper and two PF pre-impregnated kraft papers. The impregnation of the overlay paper was performed manually using a No. 9 Tan meter bar (RK Print Coat Instruments, Hertfordshire, UK). The impregnation resin was prepared by adding a curing agent (0.6 wt %), wetting agent (0.1 wt %), and slip agent (0.1 wt %) to a MF pre-condensate. Alumina particles (functionalized or non-functionalized) were added in two different concentrations: 5 and 10 wt %. A reference laminate without particles was prepared for comparison. After impregnation, the coatings were allowed to dry at room temperature for three to four days before being pressed for 1 h at 115 °C and 7.35 MPa using a Fontijne press LPB 300 apparatus (Fontijne Presses, Barendrecht, The Netherlands).

The cross-sections of selected laminates were investigated by SEM using a Hitachi SU 3500 (Hitachi High-Technologies Corporation, Minatoku, Tokyo, Japan) apparatus. Sections of approximately 2–3 cm^2 from each sample were embedded in epoxy resin under vacuum. The prepared blocks were dried at 45 °C to cure the epoxy, then ground with SiC paper and polished with diamond particles with 9 and 1 μm sizes. The samples were coated with carbon prior to SEM investigation. The images were acquired in backscatter electron imaging mode (BEI) with an accelerating voltage of 5.00 kV. The surface morphology of the scratched laminates was also investigated by SEM. For this, small sections of approximately 1 cm^2 were cut and sputter coated with a thin layer of gold. The images were acquired in secondary electron mode (SE) with an accelerating voltage of 5.00 kV.

The scratch resistance of the laminates was tested using an Anton Paar Tritec Micro-Combi Tester (Anton Paar GmbH, Graz, Austria) equipped with a Rockwell C-type diamond indenter, with a spherical tip and a radius of 100 μm. Two types of scratch tests were carried out where the load was applied either at constant mode (6 N) or at progressive mode (0–10 N). A constant scratch rate of 1 mm/min was applied for the constant load measurements and 1.66 mm/min for the progressive load measurements. The scratch profile length was 5–6 mm. At least three scratches were performed on two different replicates of each sample and the penetration depth (P_d) and the residual scratch depth (R_d) were recorded. The optical critical load, scratch width, and morphology were determined by means of optical microscopy and SEM. The elastic recovery f (%) was obtained using Equation (1) [2]:

$$f = [(P_d - R_d) \cdot 100]/P_d \tag{1}$$

Scratch resistance in terms of the residual scratch deformation of the scratch width was also taken into consideration. The scratch width was calculated as the average of the width data measured from the SEM images of the scratch tracks. The scratch hardness, defined as the normal load of the indenter over the load bearing area, was calculated using Equation (2) [28]:

$$H_s = 4qP/\pi w^2 \tag{2}$$

where w is the post-scratch width in millimeters; P is the normal load in Newtons; and q is a dimensionless parameter ($1 < q < 2$) that depends on the extent of elastic recovery of the polymer during scratching. Most commonly, for polymeric materials, the value of q is considered to be 1 [29]. For this study, the value of q was considered to be 1.

The optical critical load, L_c, defined as the normal load at which the first damage fulfilling contrast, size, and continuity criteria at the same time occurs [30], was determined. The resistance to scratch was also determined using a Taber Scratch tester model 551 (230 V, 50 Hz) in accordance with NS-EN 438-2 [31]. During this test, laminates were cut into 10-cm diameter discs and secured on the instrument turntable, then successively increasing loads (starting with 1 N) were applied and the scratches were examined visually. The scratch resistance was determined as the minimum load necessary to produce a continuous scratch mark visible to the naked eye from different angles.

3. Results

3.1. Thermogravimetric Analysis of Functionalized Alumina

TGA of the as-received alumina particles and the functionalized alumina particles is shown in Figure 2. The change in mass indicates the mass loss of the particles as the temperature increased. The functionalized particles showed a significant two-step mass loss when compared to the as-received particles, proving the existence of the silane coupling agent, which decomposes at higher temperatures. The mass loss between 50 and 140 °C corresponding to the evaporation of adsorbed moisture and water resulted from the condensation between adjacent silane agents and was not taken into account when calculating the mass loss. The mass loss between 230 and 450 °C corresponded to the covalently bound coupling agent. The mass loss was ~1.7 wt % (measured from the plateau at ~150 °C).

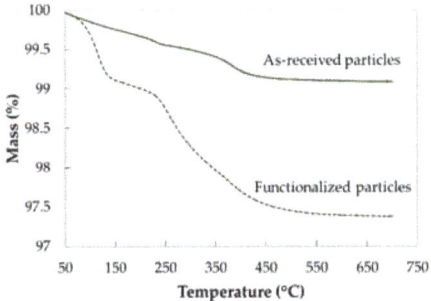

Figure 2. Mass (%) as a function of temperature of the as-received (solid line) and the functionalized (dotted line) alumina particles.

3.2. Cross-Sectional Analysis of the Laminates

The effect of the surface modification on the dispersion of the alumina particles within the resin was observed through SEM of the cross-section of the laminates, and the images are shown in Figure 3.

The alumina particles were homogeneously distributed after functionalization when compared to the as-received particles.

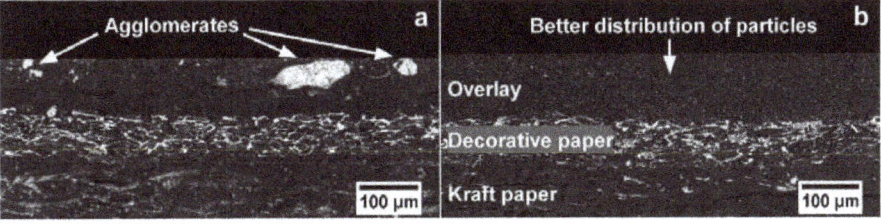

Figure 3. Cross-sectional SEM images of laminates with (**a**) 10 wt % of as-received alumina particles and (**b**) 10 wt % of functionalized alumina particles.

3.3. Scratch Resistance Measured in Accordance with Standard NS-EN 438-2

The scratch resistance was improved using the surface functionalized alumina particles. The scratch resistance results obtained through a visual inspection of the laminates after the scratch test performed in accordance with the NS-EN 438-2 standard are presented in Table 1. Using the SEM images of the scratch tracks produced through micro scratch testing, the scratch width was calculated as the average of the width data measured from the images. The values are also reported in Table 1.

The best performance was observed for the laminates containing 5 and 10 wt % of functionalized particles which were ranked with 5/6 and 6, respectively. The sample with the higher content of functionalized alumina particles showed better scratch resistance. For the laminates containing as-received particles, the opposite trend was observed, as the sample containing the higher amount of particles (10 wt %) proved to be less scratch resistant than the one containing 5 wt %. This could be due to the formation of agglomerates in the laminate with a greater amount of non-functionalized particles.

Table 1. Scratch resistance (according to standard NS-EN 438-2) and scratch widths (from micro scratch testing) of laminates containing as-received and functionalized alumina particles.

Alumina Particle Concentration (wt %)	Functionalization	Scratch Resistance (N)	Scratch Width (μm)
0	–	4	131 ± 5
5	No	5	118 ± 3
10	No	4	123 ± 6
5	Yes	5/6	120 ± 5
10	Yes	6	108 ± 7

3.4. Scratch Resistance Measured through Micro Scratch Testing

Figure 4 shows the extended scratch test results of the laminates using a micro scratch tester. Figure 4a shows the residual depth curves after a constant load of 6 N as a function of the length of the scratch and Figure 4b shows the residual scratch depth after progressive load testing from 1 to 10 N as a function of the scratch length.

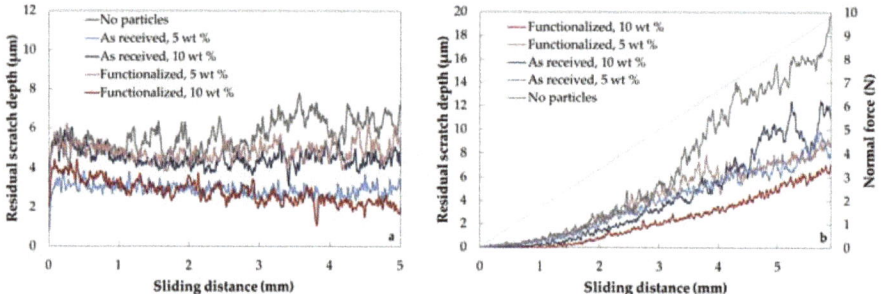

Figure 4. Residual depth after the scratch test of the laminates: (**a**) with a constant load of 6 N as a function of scratch length, and (**b**) with progressive load from 0 to 10 N as a function of scratch length.

From Figure 4a, the best performing laminate was the one containing 10 wt % of functionalized alumina particles, showing the lowest scratch depth. Compared to the laminate with 10 wt % of as-received particles, the scratch depth was more than halved from the original value (from an average 5 μm for the last two mm for the non-functionalized to 2.1 μm for the functionalized). Differences were also observed for the 5 wt % of load, however, the laminate with the as-received particles performed better, which shows that the amount of incorporated alumina particles is important. The worst performing laminate was the one containing no particles, confirming that the presence of ceramic particles is crucial.

Figure 4b shows the residual scratch depth curves after progressive load testing and confirmed that laminates with functionalized particles had the best performance. The laminates with 10 wt % of functionalized alumina were significantly better and this was especially visible for loads above 5 N. For the 5 wt % of solid loading, there was no significant difference. On the other hand, the laminate without particles showed the far largest depths, with a more jagged depth profile as a function of normal load, reflecting a higher roughness of the scratch track for this sample due to brittle fracture within the scratch track.

3.5. Scratch Microstructure and Scratch Behavior

The measured scratch resistance must be seen in conjunction with the changes in microstructure and scratch behavior. Figure 5 shows the SEM images of the residual scratch patterns of the laminates after micro scratch tests with a 6 N load.

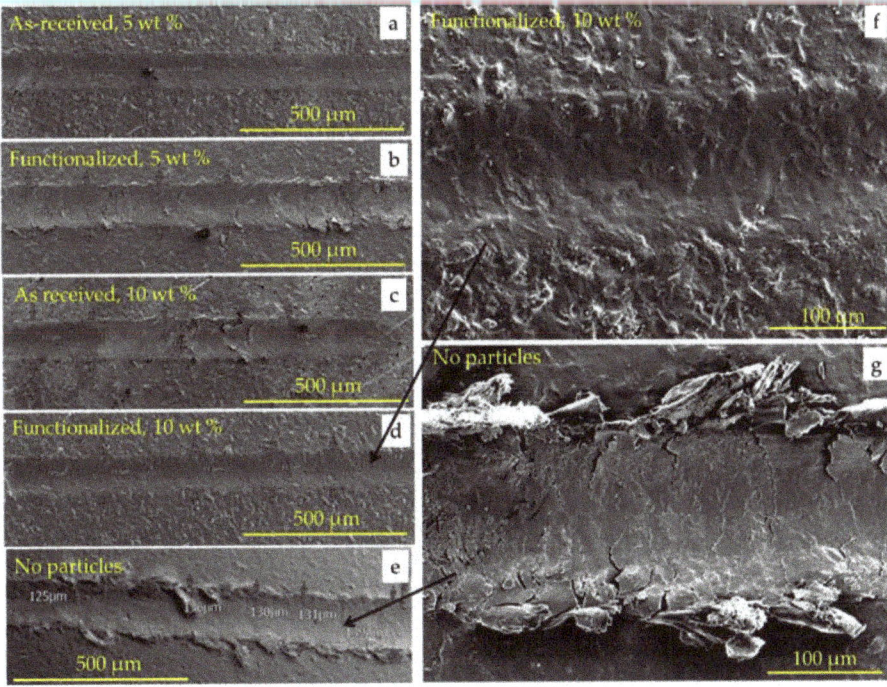

Figure 5. SEM images of laminate surfaces after the micro scratch test with a constant load of 6 N. Laminates containing (**a**) 5 wt % of as-received particles; (**b**) 5 wt % functionalized particles; (**c**) 10 wt % of as-received particles; (**d**) 10 wt % of functionalized particles, and (**e**) no particles; (**f**,**g**) 10 wt % of functionalized particles and no particles, respectively, with higher magnification.

SEM images in Figure 5f,g show the appearance of the laminate containing 10 wt % functionalized particles and of the laminate with neat polymer in the middle of the scratch. It is clear that with the functionalized alumina particles, there was no additional damage to the laminate except for the scratch itself. In the case of the laminate without particles, the damage was considerable. Here, the material was pushed upwards at the edges of the scratch and in front of the indenter and micro-cracks and plastically deformed sheet-like debris were observed in the scratch zone for the sample without particles.

The results from the scratch width measurements are shown in Table 1. The results show that for the 10 wt % of particles added, the scratches were narrower with functionalized particles when compared to the as-received particles. This confirmed that the functionalization ensured stronger bonds and yielded narrower scratches. For 5 wt %, the results were approximately the same with or without functionalization, stating that the amount of particles was important.

The post-micro scratch test elastic recovery of the laminates is presented in Figure 6, Figure 6a shows the constant load of 10 N and Figure 6b shows the progressive load from 0 to 10 N. The recovery/healing properties were high in general and were the highest for the laminates with functionalized particles. For the laminates containing 10 wt % of functionalized alumina particles, the recovery was above 80% at the end of the scratch.

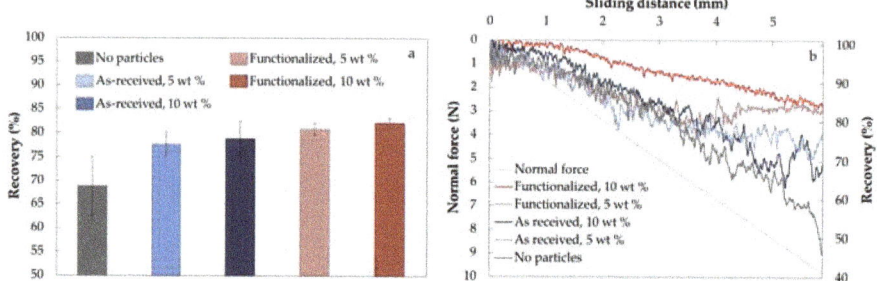

Figure 6. Elastic recovery properties of laminates after micro scratch test with (**a**) a constant load of 10 N, and (**b**) progressive load from 0 to 10 N as a function of sliding distance.

In Figure 7a, the scratch hardness calculated using the post-scratch width as well as the scratch resistance obtained according to standard BS-EN 438-2 [31] are shown. There was a good correlation between the two methods for evaluating the scratch properties. Figure 7b shows the critical loads for the visibility of the scratches obtained for the laminates. The laminates with functionalized alumina particles showed the highest critical loads, confirming their high ability to withstand high loads. These results confirmed good reproducibility between the different methods of evaluating the scratch resistance.

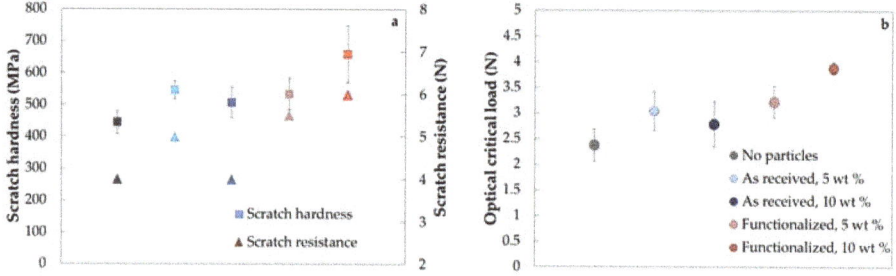

Figure 7. Scratch resistance of the laminates in terms of (**a**) scratch hardness and scratch resistance according to standard BS-EN 438-2, and (**b**) critical loads for the onset of scratch visibility in laminates containing different types of alumina particles.

4. Discussion

The scratch resistance of polymer composites depends mainly on the nature of the polymer, the additives, and the test conditions such as the applied load, the type of indenter, the scratching angle, and speed [32,33]. The complexity of the scratch behavior of polymers is due to their viscoplastic and viscoelastic nature [34], and the presence of additives such as reinforcing particles further complicates the material's response to scratching due to the formation of voids and cracks and the exposure of the particles during the scratching process [35]. In this work, functionalization of the reinforcing particles was shown to improve the scratch resistance of the laminates as well as the particle dispersion within the polymer matrix. Figure 3 clearly shows that the functionalization improved the distribution of the alumina particles within the resin. The good dispersion of the alumina particles in the laminate containing functionalized particles was due to the methoxy groups in the silane coupling agents, which contribute to the steric stabilization of the particles in the liquid phase as the polymeric chains will repulse each other. This is crucial for the performance of the laminates. It has previously been explained that while soft particles (such as agglomerates) can reduce the scratch visibility, they can also reduce the scratch hardness [12] and decrease the strength of the composite as they can act as critical flaws, especially in rigid thermosets [36] such as MF. The formation of agglomerates within the polymer matrix is therefore undesirable, and the chemical modification of the surface of the ceramic particles renders them compatible with the polymer matrix and eliminates the formation of the agglomerates.

The scratch resistance of the laminates was also significantly improved by adding functionalized particles. When comparing the results obtained from micro scratch testing, the laminate containing 10 wt % of functionalized particles gave the best results for the constant load measurements. The same observation could be made for the progressive load tests where the laminate containing 10 wt % of functionalized particles showed the best properties by far. In addition, the shapes of the residual scratch depth curves exhibited surface roughening within the scratch track for all laminates, but to a smaller extent for the one with the highest concentration of functionalized particles. This is, as was explained in a previous study [37], due to a more efficient load transfer from the matrix to the particles during the scratch test. Hence, the separation between the two phases occurred less easily and, as a result, the scratch depth decreased and the scratch resistance increased [37]. Significant differences were also seen in both the deformation extent and the deformation mechanisms when examining the scratches through microstructural investigation (Figure 5). For the neat polymer coating, the SEM images showed chipping of the surface after the scratch test all along the scratch track. However, no damage by material removal was seen on the laminate containing 10 wt % of functionalized particles. There seemed to be two types of deformation in the two cases: fracture-type and plastic-type, respectively [38]. By adding the functionalized particles, there was therefore a transition in the scratch feature. On one hand, the groove surface of the neat polymer was wave-like along the sliding direction and the edges of the scratch were sharp and irregular, with discontinuous margins and significant chip formation. The deformation included both grooving and cracking, with extensive brittle fracture both within the scratch path and on the edges. The surface texture of the scratch was also different from the texture of the unscratched coating. The difference in texture was less marked for the laminate containing 10 wt % of functionalized particles. Here, the overall features of the deformation, i.e., the smooth and continuous borders, suggest a ductile type of failure. In this case, only a small amount of material was pushed upwards at the edges of the scratch, rendering the scratch less visible. The functionalized particles were clearly reinforcing the system as less damage was seen from the scratch testing.

Higher recovery and higher scratch hardness were also obtained with particle functionalization. When recovery takes place, the bottom of the scratch moves upwards [39], and the residual depth R_d is much shallower than the penetration depth P_d. The more elastic the recovery, the smaller the scratch visibility, and these results correlated well with the visual observation results shown in Figure 7. The correlation between scratch morphology and scratch visibility is however very intricate, as it depends on may factors such as sample color, illumination, angle and duration of inspection, psychological perception as well as the size and type of damage [38]. Here, the large recovery in the scratch depth seems to be favored by the presence of the functionalized particles and it is likely that this decreased the brittleness of the polymer. The parameters influencing the recovery appear to be complex and the addition of particles can contribute to the recovery by increasing the free volume in the composite [40,41]. The phenomenon has been observed, for example, when adding silica particles, both micro- and nano-sized, to acrylic-based polyurethane coatings [41]. In this work, both non-functionalized and functionalized particles contributed to the increase in elastic properties and therefore the scratch resistance, but the functionalized particles did so to a larger extent. The same conclusion could be reached when examining the values for the critical loads, which were also higher for the laminate with a high content of functionalized alumina particles.

The enhanced scratch resistance could also be associated with a decrease in the mobility of the polymer molecules near the interface, under the influence of the fillers. The scratch resistance of the polymers and polymer composites has indeed been reported to increase with an increase in the glass transition temperature [42,43]. The enhanced glass transition temperature of polymeric composites is due to a decrease in the mobility of the polymer molecules near the interface and depends on the degree of interaction between the filler material and the polymer as well as on the interfacial area [44]. The increase in scratch resistance when adding the functionalized alumina particles to the polymer matrix can thus be attributed to an increase in the glass transition temperature and to the reduced mobility of the polymer molecules and also to the stronger interactions within the resin.

For decorative laminates, the scratch visibility is crucial and is related to the damage mechanisms and to the scratch dimensions and morphology. Brittle fractures will increase the roughness of the scratch path, chipping, shoulder formation, and scratch width. All these factors significantly affect the scratch visibility. Adding functionalized alumina particles to the polymer will decrease the scratch visibility by influencing the scratch deformation mode and, despite the subjectivity of the visual observation method used in the laminate industry to assess scratch resistance, the results correlated well with the results obtained with micro scratch testing and with the ones obtained through SEM observation.

5. Conclusions

The addition of alumina particles increased the scratch resistance of the laminates. Functionalization increased the scratch resistance further, with the best results obtained with a 10 wt % of functionalized alumina particles. An improved distribution of the particles within the resin was obtained, and the scratch resistance improvement achieved for the laminates was clearly dependent on the content and dispersion state of the particles.

Depth of the scratches and morphologies of the scratch tracks provided an insight into the deformation mode. The shape of the scratch tracks gave an indication as to the severity of the deformation as well as the scratching mechanism. By adding functionalized alumina particles, the scratch surface damage changed from extensive cracking and tearing to ductile plowing with no cracks inside the scratch track. The high scratch recovery had a great influence on the scratch visibility. Scratch recovery of the neat polymer in this study was relatively high and the addition of the functionalized alumina led to an increase in the scratch resistance and scratch hardness, but also to a reduction in scratch visibility since it further improved the recovery characteristics of the coating.

Author Contributions: Conceptualization, C.L.R., H.L.L., M.-A.E., M.B., and G.C.-C.; Methodology, C.L.R., H.L.L., M.-A.E., M.B., and G.C.-C.; Validation, C.L.R., H.L.L., M.-A.E., M.B., and G.C.-C.; Formal Analysis, C.L.R.; Investigation, C.L.R.; Resources, C.L.R., H.L.L., M.-A.E., M.B., and G.C.-C.; Data Curation, C.L.R. and T.I.H.; Writing-Original Draft Preparation, C.L.R.; Writing-Review & Editing, H.L.L., M.-A.E., M.B., and G.C.-C.; Visualization, C.L.R.; Supervision, H.L.L., M.-A.E., M.B., and G.C.-C.; Project Administration, L.K.H. and H.L.L.; Funding Acquisition, L.K.H.

Funding: This research was funded by the Research Council of Norway; the Elephant Floor project, Grant no. 228644.

Acknowledgments: We deeply acknowledge Per Olav Johnsen for his help and expertise on the SEM experiments.

Conflicts of Interest: The authors declare no conflict of interest.

References

1. Lepedat, K.; Wagner, R.; Lang, J. Laminates. In *Phenolic Resins: A Century of Progress*; Springer: Berlin/Heidelberg, Germany, 2010; pp. 243–261.
2. Krupička, A.; Johansson, M.; Hult, A. Use and interpretation of scratch tests on ductile polymer coatings. *Prog. Org. Coat.* **2003**, *46*, 32–48. [CrossRef]

3. Hara, Y.; Mori, T.; Fujitani, T. Relationship between viscoelasticity and scratch morphology of coating films. *Prog. Org. Coat.* **2000**, *40*, 39–47. [CrossRef]
4. Sobhani, H.; Khorasani, M.M. Optimization of scratch resistance and mechanical properties in wollastonite-reinforced polypropylene copolymers. *Polym. Adv. Technol.* **2016**, *27*, 765–773. [CrossRef]
5. Nemli, G. Factors affecting some quality properties of the decorative surface overlays. *J. Mater. Process. Technol.* **2008**, *195*, 218–223. [CrossRef]
6. Hossain, M.M.; Jiang, H.; Sue, H.-J. Effect of constitutive behavior on scratch visibility resistance of polymers—A finite element method parametric study. *Wear* **2011**, *270*, 751–759. [CrossRef]
7. Browning, R.L.; Jiang, H.; Sue, H.-J. Scratch behavior of polymeric materials. In *Tribology of Polymeric Nanocomposites: Friction and Wear of Bulk Materials and Coatings*, 2nd ed.; Friedrich, K., Schlarb, A.K., Eds.; Butterworth-Heinemann: Oxford, UK, 2013; pp. 513–550.
8. Brostow, W.; Cassidy, P.E.; Macossay, J.; Pietkiewicz, D.; Venumbaka, S. Connection of surface tension with multiple tribological properties in epoxy + fluoropolymer systems. *Polym. Int.* **2003**, *52*, 1498–1505. [CrossRef]
9. Xiang, C.; Sue, H.J.; Chu, J.; Coleman, B. Scratch behavior and material property relationship in polymers. *J. Polym. Sci.* **2000**, *39*, 47–59. [CrossRef]
10. Jin, P.W.; Benca, K.R.; Quarmby, I.C.; Kurpiewski, T.; Ferrell, V.E. Enhanced Scratch Resistant Coatings Using Inorganic Fillers. U.S. Patent 6,844,374, 18 January 2005.
11. Barna, E.; Bommer, B.; Kürsteiner, J.; Vital, A.; Trzebiatowski, O.V.; Koch, W.; Schmid, B.; Graule, T. Innovative, scratch proof nanocomposites for clear coatings. *Compos. Part A* **2005**, *36*, 473–480. [CrossRef]
12. Kurkcu, P.; Andena, L.; Pavan, A. An experimental investigation of the scratch behaviour of polymers—2: Influence of hard or soft fillers. *Wear* **2014**, *317*, 277–290. [CrossRef]
13. Antunes, P.V.; Ramalho, A.; Carrilho, E.V.P. Mechanical and wear behaviours of nano and microfilled polymeric composite: Effect of filler fraction and size. *Mater. Des.* **2014**, *61*, 50–60. [CrossRef]
14. Chauhan, S.R.; Thakur, S. Effects of particle size, particle loading and sliding distance on the friction and wear properties of cenosphere particulate filled vinylester composites. *Mater. Des.* **2013**, *51*, 398–408. [CrossRef]
15. Niezgoda, S.; Gupta, V.; Knight, R.; Cairncross, R.A.; Twardowski, T.E. Effect of reinforcement size on the scratch resistance and crystallinity of HVOF sprayed nylon-11/ceramic composite coatings. *J. Therm. Spray Technol.* **2006**, *15*, 731–738. [CrossRef]
16. Farzaneh, S.; Tcharkhtchi, A. Viscoelastic properties of polypropylene reinforced with mica in T_α and $T_{\alpha c}$ transition zones. *Int. J. Polym. Sci.* **2011**, *2011*, 427095. [CrossRef]
17. Sangermano, M.; Messori, M. Scratch resistance enhancement of polymer coatings. *Macromol. Mater. Eng.* **2010**, *295*, 603–612. [CrossRef]
18. Cayton, R.H.; Brotzman, R.W. Nanocomposite coatings—Applications and properties. *Mater. Res. Soc. Symp. Proc.* **2011**, *703*, V8.1. [CrossRef]
19. Wang, Y.; Lim, S.; Luo, J.L.; Xu, Z.H. Tribological and corrosion behaviors of Al_2O_3/polymer nanocomposite coatings. *Wear* **2006**, *260*, 976–983. [CrossRef]
20. Bauer, F.; Gläsel, H.-J.; Decker, U.; Ernst, H.; Freyer, A.; Hartmann, E.; Sauerland, V.; Mehnert, R. Trialkoxysilane grafting onto nanoparticles for the preparation of clear coat polyacrylate systems with excellent scratch performance. *Prog. Org. Coat.* **2003**, *47*, 147–153. [CrossRef]
21. Ye, S.; Azarnoush, S.; Smith, I.R.; Cramer, N.B.; Stansbury, J.W.; Bowman, C.N. Using hyperbranched oligomer functionalized glass fillers to reduce shrinkage stress. *Dent. Mater.* **2012**, *28*, 1004–1011. [CrossRef] [PubMed]
22. Mallakpour, S.; Madani, M. A review of current coupling agents for modification of metal oxide nanoparticles. *Prog. Org. Coat.* **2015**, *86*, 194–207. [CrossRef]
23. Ambrósio, J.D.; Balarim, C.V.M.; de Carvalho, G.B. Preparation, characterization, and mechanical/tribological properties of polyamide 11/Titanium dioxide nanocomposites. *Polym. Compos.* **2016**, *37*, 1415–1424. [CrossRef]
24. Barna, E.; Rentsch, D.; Bommer, B.; Vital, A.; Trzebiatowski, O.V.; Graule, T. Surface modification of nanoparticles for scratch resistant clear coatings. *Kautsch. Gummi Kunstst.* **2007**, *60*, 49–51.
25. Solvent Stabilizer Systems. 2017. Available online: https://www.sigmaaldrich.com/chemistry/solvents/learning-center/stabilizer-systems.html (accessed on 14 December 2017).

26. Ap-silane 51, 2015. Available online: https://static1.squarespace.com/static/57a6c9e1440243dad487f1d6/t/57b32c5746c3c465f616adbc/1471360088338/AP-SILANE+51+-+TDS.pdf (accessed on 30 March 2018).
27. Guo, Z.; Pereira, T.; Choi, O.; Wang, Y.; Hahn, H.T. Surface functionalized alumina nanoparticle filled polymeric nanocomposites with enhanced mechanical properties. *J. Mater. Chem.* **2006**, *16*, 2800–2808. [CrossRef]
28. ASTM D7027-13 Standard Test Method for Evaluation of Scratch Resistance of Polymeric Coatings and Plastics Using an Instrumented Scratch Machine; ASTM International: West Conshohocken, PA, USA, 2013.
29. Bucaille, J.L.; Felder, E.; Hochstetter, G. Mechanical analysis of the scratch test on elastic and perfectly plastic materials with the three-dimensional finite element modeling. *Wear* **2001**, *249*, 422–432. [CrossRef]
30. Rangarajan, P.; Sinha, M.; Watkins, V.; Harding, K.; Sparks, J. Scratch visibility of polymers measured using optical imaging. *Polym. Eng. Sci.* **2003**, *43*, 749–758. [CrossRef]
31. BS EN 438-2:2016 High-Pressure Decorative Laminates (HPL). Sheets Based on Thermosetting Resins (Usually Called Laminates). Part 2: Determination of Properties; British Standards Institution: London, UK, 2016.
32. Rajesh, J.J.; Bijwe, J. Investigations on scratch behaviour of various polyamides. *Wear* **2005**, *259*, 661–668. [CrossRef]
33. Briscoe, B.J.; Sinha, S.K. Scratch resistance and localised damage characteristics of polymer surfaces—A review. *Materialwiss. Werkstofftech.* **2003**, *34*, 989–1002. [CrossRef]
34. Al-Rub, R.K.A.; Tehrani, A.H.; Darabi, M.K. Application of a large deformation nonlinear-viscoelastic viscoplastic viscodamage constitutive model to polymers and their composites. *Int. J. Damage Mech.* **2015**, *24*, 198–244. [CrossRef]
35. Wong, M.; Moyse, A.; Lee, F.; Sue, H.-J. Study of surface damage of polypropylene under progressive loading. *J. Mater. Sci.* **2004**, *39*, 3293–3308. [CrossRef]
36. Rothon, R.N. *Particulate Fillers for Polymers*; Rapra review reports, Vol. 12; Smithers Rapra Publishing: Shrewsbury, UK, 2002.
37. Khalilnezhad, P.; Sajjadi, S.A.; Zebarjad, S.M. Effect of nanodiamond surface functionalization using oleylamine on the scratch behavior of polyacrylic/nanodiamond nanocomposite. *Diam. Relat. Mater.* **2014**, *45*, 7–11. [CrossRef]
38. Lin, L.; Blackman, G.S.; Matheson, R.R. A new approach to characterize scratch and mar resistance of automotive coatings. *Prog. Org. Coat.* **2000**, *40*, 85–91. [CrossRef]
39. Brostow, W.; Deborde, J.-L.; Jaklewicz, M.; Olszynski, P. Tribology with emphasis on polymers: Friction, scratch resistance and wear. *J. Mater. Educ.* **2003**, *24*, 119–132.
40. Brostow, W.; Hagg Lobland, H.E.; Narkis, M. Sliding wear, viscoelasticity, and brittleness of polymers. *J. Mater. Res.* **2006**, *21*, 2422–2428. [CrossRef]
41. Zhou, S.; Wu, L.; Sun, J.; Shen, W. The change of the properties of acrylic-based polyurethane via addition of nano-silica. *Prog. Org. Coat.* **2002**, *45*, 33–42. [CrossRef]
42. Lange, J.; Luisier, A.; Hult, A. Influence of crosslink density, glass transition temperature and addition of pigment and wax on the scratch resistance of an epoxy coating. *J. Coat. Technol.* **1997**, *69*, 77–82. [CrossRef]
43. Rink, H.-P. Polymeric engineering for automotive coating applications. In *Automotive Paints and Coatings*; Streitberger, H.-J., Kreis, W., Eds.; Wiley-VCH Verlag GmbH & Co. KGaA: Chichester, UK, 2008; pp. 211–257.
44. Droste, D.H.; Dibenedetto, A.T. The glass transition temperature of filled polymers and its effect on their physical properties. *J. Appl. Polym. Sci.* **1969**, *13*, 2149–2168. [CrossRef]

© 2018 by the authors. Licensee MDPI, Basel, Switzerland. This article is an open access article distributed under the terms and conditions of the Creative Commons Attribution (CC BY) license (http://creativecommons.org/licenses/by/4.0/).

Article

A Ladder-Type Organosilicate Copolymer Gate Dielectric Materials for Organic Thin-Film Transistors

Dongkyu Kim and Choongik Kim *

Department of Chemical and Biomolecular Engineering, Sogang University, Seoul 04107, Korea; dgkim1993@gmail.com
* Correspondence: choongik@sogang.ac.kr; Tel.: +82-2-705-7964; Fax: +82-2-711-0439

Received: 31 May 2018; Accepted: 2 July 2018; Published: 3 July 2018

Abstract: A ladder-type organosilicate copolymer based on trimethoxymethylsilane (MTMS) and 1,2-bis(triethoxysilyl)alkane (BTESn: n = 2–4) were synthesized for use as gate dielectrics in organic thin-film transistors (OTFTs). For the BTESn, the number of carbon chains (2–4) was varied to elucidate the relationship between the chemical structure of the monomer and the resulting dielectric properties. The developed copolymer films require a low curing temperature ($\approx$150 °C) and exhibit good insulating properties (leakage current density of $\approx 10^{-8}$–10^{-7} A·cm^{-2} at 1 MV·cm^{-1}). Copolymer films were employed as dielectric materials for use in top-contact/bottom-gate organic thin-film transistors and the resulting devices exhibited decent electrical performance for both p- and n-channel organic semiconductors with mobility as high as 0.15 cm^2·V^{-1}·s^{-1} and an I_{on}/I_{off} of >10^5. Furthermore, dielectric films were used for the fabrication of TFTs on flexible substrates.

Keywords: organic thin-film transistors; dielectric; organosilicate; copolymer

1. Introduction

Organic thin-film transistors (OTFTs) have been an important research subject because they are indispensable elements in the development of low-cost, large-area electronics, such as paper-based displays, smart cards, radio-frequency ID tags, and sensors [1–8]. Among the fundamental components of OTFTs, namely the semiconductor, dielectric, and conductor, studies on the development and application of organic semiconducting materials have intensively been performed for the last few decades [9–18]. Development of new dielectric materials for OTFTs, which perform as capacitors, insulators, and substrates, is also important to realize various electronic applications based on OTFTs. Favorable dielectric characteristics of OTFT-based electronic devices include ease of preparation, low cost, large area processibility, as well as excellent insulating properties [19–23].

Inorganic materials, such as SiO$_2$, Al$_2$O$_3$, and TiO$_2$, have traditionally been employed as dielectrics for OTFTs due to their excellent insulating properties. However, inorganic dielectric materials require a high-temperature/high-cost vacuum process for their preparation [24,25]. On the other hand, organic dielectric materials, such as polymers, can readily be processed from solution at a relatively low temperature over a large area, while their insulating properties are inferior to those of inorganic counterparts [26–28]. Furthermore, some organic-based dielectrics can be readily processed at low temperature, which can be employed on flexible substrates, such as polyethylene terephthalate (PET) and polyethylene naphthalate (PEN) [29–38].

To this end, many studies on the development and application of new dielectric materials with low leakage current, high dielectric constant, and facile process have been reported [39–48]. Among them, organic-inorganic hybrid dielectrics might synergistically combine the advantages of both materials, i.e., the excellent mechanical and electrical properties of inorganics and the flexibility and large area solution processibility of organics [41–44]. One of the representative examples is organosilicate materials, a type of silica containing organic groups. A few studies employing dielectric materials containing organosilicates have been reported for the fabrication of OTFTs [45–48].

For instance, Kang et al. reported silsesquioxane copolymer gate dielectrics based on methacryloxy propyltrimethoxysilane and phenyltrimethoxysilane for use in OTFTs. The reported dielectric films formed at a curing temperature of 200 °C exhibited a leakage current density as low as 6×10^{-9} A·cm^{-2} at 1 MV·cm^{-1} [45]. Similarly, Kim et al. reported dielectric materials based on a silsesquioxane derivative containing cage-structured epoxy groups, and the resulting films showed a leakage current density of $\approx 10^{-8}$ A·cm^{-2} at 1 MV·cm^{-1} [46]. Lee et al. reported hybrid copolymer gate dielectrics based on methyltrimethoxysilane and 1,2-bis(trimethoxysilyl)ethane for use in OTFTs, with a leakage current density of 2.9×10^{-6} A·cm^{-2} at 0.5 MV·cm^{-1} [47]. Similarly, Hamada et al. reported on a polymethylsilsesquioxane (PMSQ) dielectric prepared by a sol-gel method from methyltrimethoxysilane. While the dielectric characteristic (leakage current density) was not reported in this study, the OTFTs based on the PMSQ dielectric showed very poor electrical performance with a mobility of ≈ 0.006 cm^2·V^{-1}·s^{-1} [48].

In this study, we report a ladder-type copolymer based on trimethoxymethylsilane (MTMS) and 1,2-bis(triethoxysilyl)alkane (BTESn) as dielectric materials. Previously reported organosilicate copolymer dielectric films based on similar monomers exhibited high leakage current density, resulting in relatively poor device characteristic for poly(3-hexylthiophene) (P3HT), possibly due to the unoptimized processing conditions [47]. Hence, we have optimized the synthesis and processing condition of copolymers by varying the monomer ratio, solvent, and employing different types of monomers. The copolymers were synthesized by a simple sol-gel reaction of two types of monomers, and curing was performed at a relatively low annealing temperature of 150 °C. The resulting films exhibited a leakage current density of 10^{-8}–10^{-7} A·cm^{-2} at 1 MV·cm^{-1} at a thickness of ≈ 100 nm. Furthermore, the correlation between the dielectric properties of the copolymers and types of monomers was investigated by varying carbon chain lengths in BTESn. The OTFTs employing copolymer dielectrics were fabricated, and the resulting devices were active for both p- and n-channel organic semiconductors with carrier mobility as high as 0.15 cm^2 V^{-1}·s^{-1} and a current on/off ratio of $\approx 10^5$ (Figure 1). Furthermore, we have employed the developed copolymer dielectric films on flexible substrate to fabricate flexible TFTs.

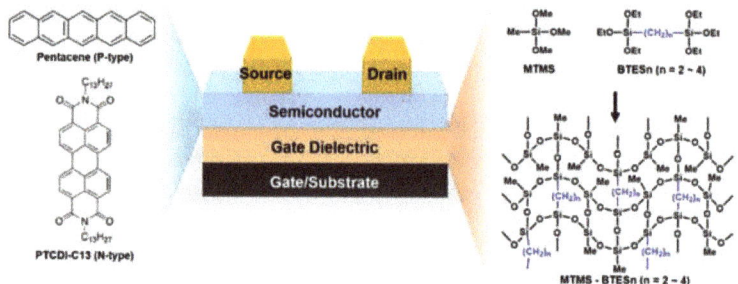

Figure 1. Schematic of top-contact/bottom-gate OTFT and chemical structures of organic semiconductors and dielectric materials employed in this study.

2. Materials and Methods

2.1. Materials and Methods

The semiconductors pentacene (P5) and N,N'-Dioctyl-3,4,9,10-perylenedicarboximide (PTCDI-C13) were purchased from Sigma-Aldrich (St. Louis, MO, USA). Trimethoxymethylsilane (MTMS) and 1,2-bis(triethoxysilyl) ethane (BTES2), 1,2-bis(triethoxysilyl) propane (BTES3), 1,2-bis(triethoxysilyl) butane (BTES4) were purchased from Sigma-Aldrich (St. Louis, MO, USA).

All materials were used without further purification. PEN substrates (Film Type Teonex Q65H, DuPont Teijin Films, Chester, VA, USA) were used as received without solvent cleaning. Structures of the MTMS-BTESn were studied by ^{1}H NMR spectroscopy (Varian UNITY-INOVA 500 MHz NMR spectrometer using Acetone-d$_6$ as a solvent, Palo Alto, CA, USA) and Fourier transform infrared spectroscopy (FT-IR, Thermo Scientific Nicolet iS50R FTIR Spectrometer, Thermo Fisher Scientific, Waltham, MA, USA). Thermogravimetric analyses (TGA, TA Instrument Q50-1555, New Castle, DE, USA) were performed on each sample in a platinum crucible from 40 to 800 °C at a heating rate of 10 °C min^{-1} under an N$_2$ atmosphere.

2.2. Synthesis

2.2.1. Synthesis of MTMS-BTES2

Trimethoxymethylsilane (5.11 g, 37.5 mmol) and 1,2-bis (triethoxysilyl) ethane (4.43 g, 12.5 mmol) were added to anhydrous tetrahydrofuran (25 mL) and stirred at room temperature for 30 min. To this solution was added a 2N aqueous hydrochloric acid solution (0.75 mL) and distilled water (8.32 mL), and the mixture was stirred at 40 °C for 6 h under a nitrogen atmosphere to perform a sol-gel reaction. The reaction mixture was cooled to room temperature, and the organosilicate precursor synthesized from the solution was extracted with diethyl ether (50 mL). The extracted precursor solution was washed twice with distilled water, and then dried under vacuum at room temperature for 1 h to remove the solvent to obtain a white powdery solid. The solids were again dissolved in acetone (200 mL) to afford the gelation of the organosilicate, and undissolved particles were removed by filtration. The solution was dried under vacuum at room temperature for 12 h using a rotary evaporator to remove residual solvent, and a white solid (3.97 g, 41.61%) was obtained again. ^{1}H-NMR (500 MHz, Acetone-d$_6$): δ 0.19 (s, 3H), 0.90 (s, 2H), 5.70 (s, H).

2.2.2. Synthesis of MTMS-BTES3

Trimethoxymethylsilane (5.11 g, 37.5 mmol) and 1,2-bis (triethoxysilyl) propane (4.61 g, 12.5 mmol) were added to anhydrous tetrahydrofuran (25 mL) and stirred at room temperature for 30 min. To this solution was added a 2N aqueous hydrochloric acid solution (0.75 mL) and distilled water (8.32 mL), and the mixture was stirred at 40 °C for 6 h under a nitrogen atmosphere to perform a sol-gel reaction. The next purification processes were same as those of MTMS-BTES2 and a white solid (4.43 g, 45.58%) was obtained. ^{1}H-NMR (500 MHz, Acetone-d$_6$): δ 0.19 (s, 3H), 0.90 (s, 2H), 1.71 (s, 2H), 5.70 (s, H).

2.2.3. Synthesis of MTMS-BTES4

Trimethoxymethylsilane (5.11 g, 37.5 mmol) and 1,2-bis (triethoxysilyl) butane (4.78 g, 12.5 mmol) were added to anhydrous tetrahydrofuran (25 mL) and stirred at room temperature for 30 min. To this solution was added a 2N aqueous hydrochloric acid solution (0.75 mL) and distilled water (8.32 mL), and the mixture was stirred at 40 °C for 6 h under a nitrogen atmosphere to perform a sol-gel reaction. The next purification processes were same as those of MTMS-BTES2 and a white solid (4.68 g, 47.32%) was obtained. ^{1}H-NMR (500MHz, Acetone-d$_6$): δ 0.19 (s, 3H), 0.90 (s, 2H), 1.71 (s, 2H), 5.70 (s, H).

2.3. Film Preparation and Device Fabrication

Top-contact/bottom-gate organic thin-film transistor devices were fabricated using highly n-doped silicon wafers (resistivity < 0.005 Ω·cm) as gate/substrates. The substrates were cleaned by sonication in isopropyl alcohol for 20 min, followed by O$_2$ plasma for 5 min (Harrick Plasma, 18W, Ithaca, NY, USA). For the fabrication of devices on flexible substrates, the bottom Al gate electrodes were thermally evaporated through a shadow mask on the polyethylene naphthalate (PEN) substrate. For the preparation of a gate dielectric layer, MTMS-BTESn (n = 2–4) was dissolved in propylene glycol monomethyl ether acetate (PGMEA) solvent at 10 wt.%. The resulting solutions were dropped on a silicon wafer or PEN substrates with Al gate electrodes through a syringe equipped with a 0.2 μm polytetrafluoroethylene (PTFE) filter, and then spun at 3000 rpm for 30 s. The spin-coated films were annealed at 150 °C for 2 h under nitrogen. Film thicknesses were 80–150 nm, as characterized by an ellipsometer. A 40 nm thick P5 or PTCDI-C13 film was deposited onto the prepared MTMS-BTESn copolymer gate dielectrics at a rate of 0.2 Å·s^{-1} ($\approx$10^{-5} Torr). Finally, a gold source and drain electrode (40 nm) was thermally evaporated through a shadow mask with channel lengths (L, 50–100 µm) and width (W, 1000–2000 µm).

Metal/insulator/semiconductor (MIS) structures were fabricated using highly n-doped silicon wafers as a gate/substrate. Then the organosilicate copolymer dielectric was spin-coated and annealed at 150 °C for 2 h, as described above. Finally, gold top electrodes (40 nm) were thermally deposited. The area of the testing capacitor was 0.2 mm × 0.2 mm.

2.4. Film and Device Characterization

Film thicknesses were characterized by ellipsometer (Filmetrics, F20, San Diego, CA, USA). The surface morphologies of copolymer thin films were characterized by atomic force microscope (AFM, NX10, Park Systems, Suwon, South Korea). A Keithley 4200 SCS (Cleveland, OH, USA) was used for the characterization of OTFTs and leakage current measurements. The capacitance measurement of copolymer thin films were performed using an LCR meter (Agilent 4284A, Agilent Technologies, Santa Clara, CA, USA). Carrier mobilities (μ) were determined in the saturation regime using the standard relationship, $\mu_{sat} = (2I_{DS}L)/[WC_i(V_G - V_T)^2]$ (I_{DS}, source-drain current; L, the channel length; W, channel width; C_i, areal capacitance of the gate dielectric; V_G, gate voltage; V_T, threshold voltage).

3. Results

3.1. Synthesis

MTMS-BTESn copolymers were prepared using a sol-gel reaction of MTMS and BTESn, and were characterized by ^{1}H-NMR and FT-IR (Figures S1 and S2). As shown in the NMR spectra (Figure S1), the MTMS monomer was confirmed by the pendant methyl group peak at $\approx$0.08–0.2 ppm. The BTESn monomer was confirmed by ethylene bridge structure peak formed at 0.8 ppm. BTES3 and BTES4 were additionally confirmed at $\approx$1.6 ppm using a bridge structure peak. Note that the peak at $\approx$6 ppm is from the residual hydroxide group formed during the polymerization. The network structure formed between MTMS and BTESn can be confirmed by FT-IR spectra (Figure S2). Two absorption peaks at 1131 and 1040 cm^{-1}, corresponding to Si–O–Si stretching vibrations, were observed in all samples. The bimodal absorption peaks indicate the formation of ladder-type structures [49–51].

3.2. Dielectric Characterization

The MTMS-BTESn copolymer film was formed via spin-coating onto a substrate to a thickness of $\approx$80–150 nm, as measured by an ellipsometer. First, the surface morphologies of the dielectric films were characterized using an AFM, and a relatively smooth surface (root-mean-square roughness: 0.4–1.0 nm) was observed, regardless of the film thickness (Figure 2).

Then, the leakage current density and areal capacitance of the dielectric films were measured based on the MIS structure with Au top electrodes (Table 1 and Figure 3). First, all dielectric films showed a low leakage current density of $\approx$10^{-8}–10^{-7} A·cm^{-2} at 1 MV·cm^{-1} (Figure 3a). Although there was no significant difference with regard to the type of monomer, the leakage current density slightly increased as the number of carbon atoms of BTESn increased (Figure 3c). The methyl groups in the MTMS might reduce the relative amount of hydroxyl groups during the sol-gel reaction and increase the stability of the film [37]. Also, electrical stability of the dielectric can be enhanced by reducing the free volume due to the low curing temperature of 150 °C and by limiting the thermal dynamic motion of the polymer chains due to the sol-gel bonded network structure.

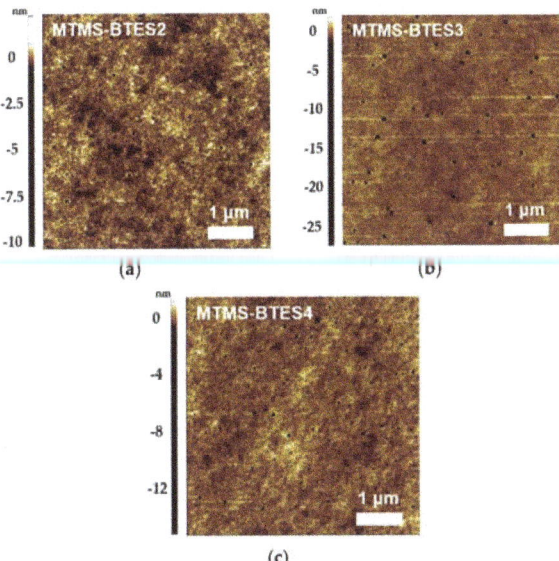

Figure 2. AFM images (5 μm × 5 μm) of MTMS-BTES*n* copolymer gate dielectric films: (a) MTMS-BTES2, (b) MTMS-BTES3, and (c) MTMS-BTES4.

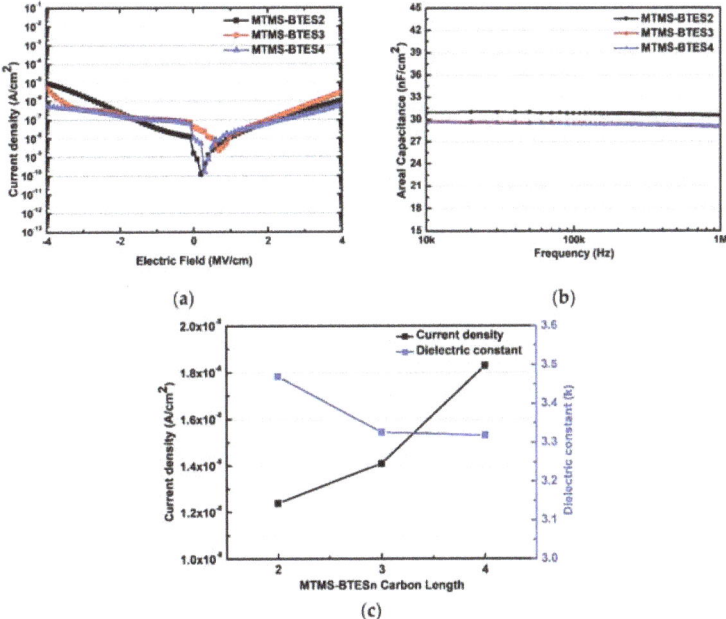

Figure 3. Electrical properties of spin-coated copolymer films fabricated with various monomers (thickness: 100 nm): (a) Leakage current density vs electric field plot; (b) Capacitance-frequency plots ($10–10^3$ kHz); (c) Comparison of leakage current density and k value as a function of monomer type.

Table 1. Properties of Copolymer Dielectrics based on MIS Device Structure (J, leakage current density; C_i, Capacitance; k, dielectric constants).

Copolymer	Thickness (nm)	J (A·cm^{-2}) at 1 MV·cm^{-1}	C_i (nF·cm^{-2})	k
MTMS-BTES2	140	8.11×10^{-8}	23.0	3.60
	100	1.24×10^{-8}	30.8	3.47
	80	1.38×10^{-7}	33.6	3.33
MTMS-BTES3	135	3.91×10^{-8}	23.3	3.50
	102	1.41×10^{-8}	29.5	3.33
	90	1.20×10^{-8}	32.2	3.27
MTMS-BTES4	150	5.27×10^{-8}	20.5	3.47
	113	6.55×10^{-8}	29.1	3.31
	100	1.83×10^{-8}	29.4	3.32

In the case of the capacitance measurement, the measurement was performed between 10 and 10^3 kHz. As the frequency increased, the capacitance value decreased slightly. Based on the capacitance value at 10^2 kHz, the dielectric constants were 3.3–3.5, which is similar to the conventional organosilicate-based materials [35–37]. For the dielectric films with comparable thickness, as the number of carbon in the monomers increased, the value of k decreased, which could be a result of the increased free volume of monomers with more carbon chains.

3.3. Thin-Film Transistor Fabrication and Characterization

We fabricated OTFTs by spin-coating MTMS-BTESn materials on the n^{++}-Si gate/substrate, followed by annealing at 150 °C for 2 h. P-type (P5) and n-type (PTCDI-C13) organic semiconductors were vacuum-deposited onto the copolymer dielectric layer, followed by Au source/drain electrodes. The device characteristics including carrier mobilities in the saturation regime (μ_{sat}), and current on/off ratios (I_{on}/I_{off}), and threshold voltage (V_T) are shown in Table 2. Transfer characteristics of the devices were measured in the saturation regime ($V_{DS} \geq V_G - V_T$), and representative transfer plots are shown in Figure 4. As shown, the OTFT devices operated well for both p- and n-channel organic semiconductors with a carrier mobility as high as 0.15 cm^2·V^{-1}·s^{-1} and I_{on}/I_{off} of >10^5. Furthermore, copolymer dielectric films could be employed on flexible PEN substrates, due to the low curing temperature, and the resulting OTFT device showed a slightly lower performance than that on a rigid substrate with an electron mobility of 0.04 cm^2·V^{-1}·s^{-1} for the PTCDI-C13 semiconductor (Figure 5).

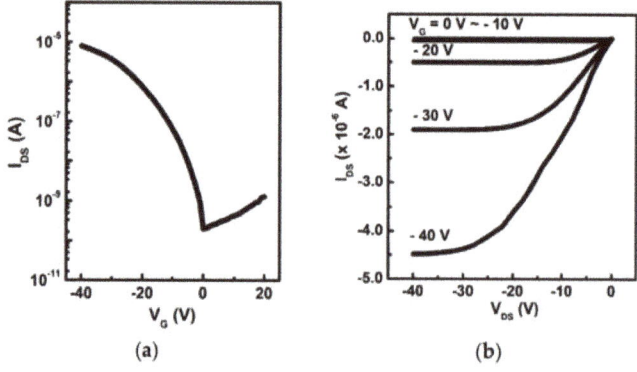

Figure 4. Cont.

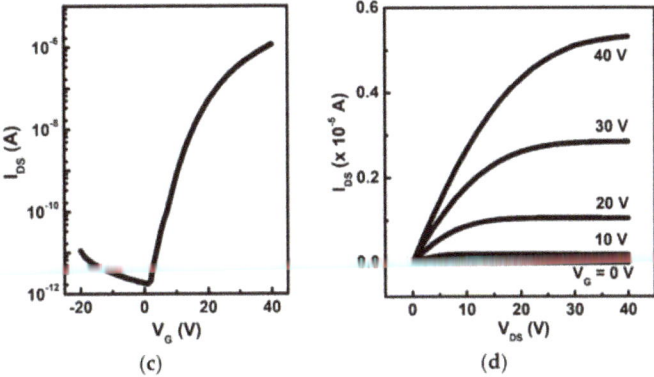

Figure 4. Performance of OTFT devices on copolymer dielectric films. OTFT transfer characteristics for (**a**) P5, (**c**) PTCDI-C13, and OTFT output characteristics for (**b**) P5, (**d**) PTCDI-C13. The channel width and length are 1000 and 100 μm, respectively.

Table 2. TFT Device Performance Parameters based on P5 and PTCDI-C13 Semiconductors on Copolymer Dielectric. (μ_{sat}: carrier mobility, I_{on}/I_{off}: current on/off ratio, V_T; threshold voltage).

Type	Semi-Conductor	μ_{sat} [1] $(cm^2 \cdot V^{-1} \cdot s^{-1})$	I_{on}/I_{off}	V_T (V)
p-type	P5	0.15	2.24×10^5	−10.5
n-type	PTCDI-C13	0.09	3.00×10^6	9.3

[1] The average values obtained from six different devices.

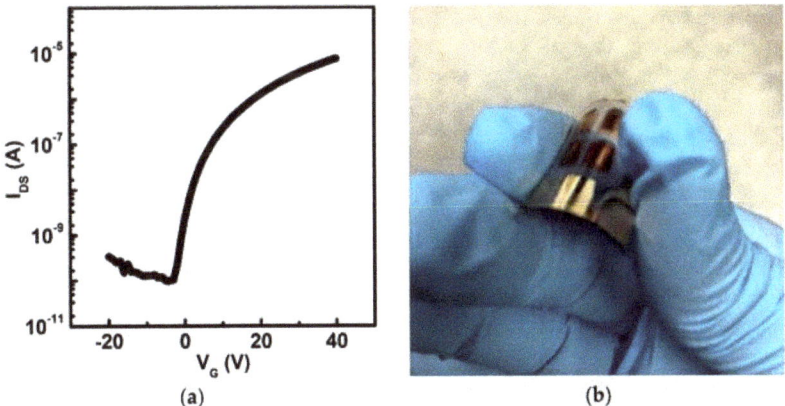

Figure 5. OTFT devices fabricated on flexible PEN substrates. (**a**) Transfer characteristics using a PTCDI-C13 as semiconductor (channel width and length are 2000 and 50 μm, respectively); (**b**) Photo of flexible OTFT device.

4. Conclusions

This article explored the dielectric characteristics of a copolymer based on MTMS and BTESn with various carbon numbers. Organosilicate copolymer dielectrics with various lengths of carbon chains could be easily prepared by the sol-gel method. Despite the existence of natural free volume due to the difference of compatibility between organic and inorganic materials, a good insulating property of copolymer dielectrics was observed. All three types of dielectric films exhibited a relatively smooth surface and low leakage current density of $\approx 10^{-8}$–10^{-7} A·cm^{-2} at 1 MV·cm^{-1} via the optimization of processing conditions and monomer components. With regard to different chemical structure of monomers, the leakage current density increased slightly, while dielectric constant value decreased, as the number of carbon atoms in BTESn increased, possibly due to the structural free volume caused by the carbon bridge. The copolymer films were employed as a dielectric layer for OTFTs, and the resulting devices were active for both p- and n-channel organic semiconductors with a carrier mobility as high as 0.15 cm^2·V^{-1}·s^{-1} and I_{on}/I_{off} of $\approx 10^5$.

Supplementary Materials: The following are available online at http://www.mdpi.com/2079-6412/8/7/236/s1, Figure S1: ^{1}H-NMR analysis of MTMS-BTESn copolymer, Figure S2: Fourier transform infrared spectroscopy (FT-ir) of MTMS-BTESn copolymer

Author Contributions: D.K. and C.K. designed the experiments. D.K. performed the experiments. D.K. and C.K. wrote the paper.

Funding: This research was funded by the National Research Foundation of Korea (NRF) (NRF-2017R1A2B4001955 and NRF-2017M2B2A4049297).

Conflicts of Interest: The authors declare no conflict of interest.

References

1. Ortiz, R.P.; Facchetti, A.; Marks, T.J. High-k Organic, Inorganic, and Hybrid Dielectrics for Low-Voltage Organic Field-Effect Transistors. *Chem. Rev.* **2010**, *110*, 205–239. [CrossRef] [PubMed]
2. Kang, B.S.; Lee, W.H.; Cho, K.W. Recent Advances in Organic Transistor Printing Processes. *ACS Appl. Mater. Interfaces* **2013**, *5*, 2302–2315. [CrossRef] [PubMed]
3. Forrest, S.B. The path to ubiquitous and low-cost organic. *Nature* **2004**, *428*, 911–918. [CrossRef] [PubMed]
4. Yi, H.T.; Payne, M.M.; Anthony, J.E.; Podzorov, V. Ultra-flexible solution-processed organic field-effect transistors electronic appliances on plastic. *Nat. Commun.* **2012**, *3*, 1259–1265. [CrossRef] [PubMed]
5. Ozdemir, M.; Choi, D.H.; Kwon, G.H.; Zorlu, Y.; Cosut, B.; Kim, H.K.; Facchetti, A.; Kim, C.I.; Usta, H. Solution-Processable BODIPY-Based Small Molecules for Semiconducting Microfibers in Organic Thin-Film Transistors. *ACS Appl. Mater. Interfaces* **2016**, *8*, 14077–14087. [CrossRef] [PubMed]
6. Sirringhaus, H. 25th Anniversary Article: Organic Field-Effect Transistors: The Path Beyond Amorphous Silicon. *Adv. Mater.* **2014**, *26*, 1319–1335. [CrossRef] [PubMed]
7. Cowen, L.M.; Atoyo, J.; Carnie, M.J.; Baran, D.; Schroeder, B.C. Review—Organic Materials for Thermoelectric Energy Generation. *ECS J. Solid State Sci. Technol.* **2017**, *6*, 3080–3088. [CrossRef]
8. Pierre, A.; Sadeghi, M.; Payne, M.M.; Facchetti, A.; Anthony, J.E.; Arias, A.C. All-Printed Flexible Organic Transistors Enabled by Surface Tension-Guided Blade Coating. *Adv. Mater.* **2014**, *26*, 5722–5727. [CrossRef] [PubMed]
9. Heitzer, H.M.; Marks, T.J.; Ratner, M.A. Molecular Donor—Bridge—Acceptor Strategies for High-Capacitance Organic Dielectric Materials *J. Am. Chem. Soc.* **2015**, *137*, 7189–7196. [CrossRef] [PubMed]
10. Yan, Y.; Huang, L.B.; Zhou, Y.; Han, S. T; Zhou, L.; Sun, Q.; Zhuang, J.; Peng, H.; Yan, H.; Roy, V.A.L. Surface Decoration on Polymeric Gate Dielectrics for Flexible Organic Field-Effect Transistors via Hydroxylation and Subsequent Monolayer Self-Assembly. *ACS Appl. Mater. Interfaces* **2015**, *7*, 23464–23471. [CrossRef] [PubMed]
11. Jiang, Y.; Guo, Y.; Liu, Y. Engineering of Amorphous Polymeric Insulators for Organic Field-Effect Transistors. *Adv. Electron. Mater.* **2017**, *3*, 1700157. [CrossRef]
12. Kim, H.K.; Reddy, M.R.; Kim, H.S.; Choi, D.H.; Kim, C.I.; Seo, S.Y. Benzothiadiazole-Based Small-Molecule Semiconductors for Organic Thin-Film Transistors and Complementary-like Inverters. *ChemPlusChem* **2017**, *82*, 742–749. [CrossRef]
13. Ozdemir, M.; Choi, D.H.; Zorlu, Y.; Cosut, B.; Kim, H.S.; Kim, C.I.; Usta, H. A new rod-shaped BODIPY-acetylene molecule for solution-processed semiconducting microribbons in n-channel organic field-effect transistors. *New J. Chem.* **2017**, *41*, 6232–6240. [CrossRef]

14. Ozdemir, R.; Choi, D.H.; Ozdemir, M.; Kim, H.K.; Kostakoglu, S.T.; Erkartal, M.; Kim, H.S.; Kim, C.I.; Usta, H. A Solution-Processable Liquid-Crystalline Semiconductor for Low-Temperature-Annealed Air-Stable N-Channel Field-Effect Transistors. *ChemPhysChem* **2017**, *18*, 850–861. [CrossRef] [PubMed]
15. Vegiraju, S.; He, G.Y.; Kim, C.I.; Priyanka, P.; Chiu, Y.J.; Liu, C.W.; Huang, C.Y.; Ni, J.S.; Wu, Y.W.; Chen, Z.; et al. Solution-Processable Dithienothiophenoquinoid (DTTQ) Structures for Ambient-Stable n-Channel Organic Field Effect Transistors. *Adv. Funct. Mater.* **2017**, *27*, 1606761. [CrossRef]
16. Ozdemir, M.; Kim, S.W.; Kim, H.S.; Kim, M.G.; Kim, B.J.; Kim, C.I.; Usta, H. Semiconducting Copolymers Based on meso-Substituted BODIPY for Inverted Organic Solar Cells and Field-Effect Transistors. *Adv. Electron. Mater.* **2017**, 1700354. [CrossRef]
17. Ho, D.I.; Jeon, M.S.; Kim, H.K.; Gidron, O.; Kim, C.I.; Seo, S.Y. Solution-processable dithieno [3,2-b:2′,3′-d] thiophene derivatives for organic thin-film transistors and complementary-like inverters. *Org. Electron.* **2018**, *52*, 356–363. [CrossRef]
18. Reddy, M.R.; Kim, H.S.; Kim, C.I.; Seo, S.Y. 2-Thiopene [1] benzothieno [3,2-b] benzothiophene derivatives as solution processable organic semiconductors for organic thin-film transistors. *Synth. Met.* **2018**, *235*, 153–159. [CrossRef]
19. Bao, Z.; Chen, X. Flexible and Stretchable Devices. *Adv. Mater.* **2016**, *18*, 4177–4179. [CrossRef] [PubMed]
20. Nela, L.; Tang, J.; Cao, Q.; Tulevski, G.; Han, S.J. Large-Area High-Performance Flexible Pressure Sensor with Carbon Nanotube Active Matrix for Electronic Skin. *Nano Lett.* **2018**, *18*, 2054–2059. [CrossRef] [PubMed]
21. Lei, T.; Pochorovski, I.; Bao, Z. Separation of Semiconducting Carbon Nanotubes for Flexible and Stretchable Electronics Using Polymer Removable Method. *Acc. Chem. Res.* **2017**, *50*, 1096–1104. [CrossRef] [PubMed]
22. Joe, D.J.; Kim, S.J.; Park, J.H.; Park, D.Y.; Lee, H.E.; Im, T.H.; Choi, I.S.; Ruoff, R.S.; Lee, K.J. Laser-Material Interactions for Flexible Applications. *Adv. Mater.* **2017**, *29*, 1606586. [CrossRef] [PubMed]
23. Myny, K. The development of flexible integrated circuits based on thin-film transistors. *Nat. Electron.* **2018**, *1*, 30–39. [CrossRef]
24. Ha, T.J.; Sonar, P.; Dodabalapur, A. Improved Performance in Diketo pyrrolopyrrole-Based Transistors with Bilayer Gate Dielectrics. *ACS Appl. Mater. Interfaces* **2014**, *6*, 3170–3175. [CrossRef] [PubMed]
25. Tardy, J.; Erouel, M.; Deman, A.L.; Gagnaire, A.; Teodorescu, V.; Blanchin, M.G.; Canut, B.; Barau, A.; Zaharescu, M. Organic thin film transistors with HfO$_2$ high-k gate dielectric grown by anodic oxidation or deposited by sol-gel. *Microelectron. Reliab.* **2007**, *47*, 372–377. [CrossRef]
26. Li, S.X.; Feng, L.R.; Guo, X.J.; Zhang, Q. Application of thermal azide-alkyne cycloaddition(TAAC) reaction as a low temperature cross-linking method in polymer gate dielectrics for organic field-effect transistors. *J. Mater. Chem. C* **2014**, *2*, 3517–3520. [CrossRef]
27. Puigdollers, J.; Voz, C.; Orpella, A.; Quidant, R.; Martin, I.; Vetter, M.; Alcubilla, R. Pentacene thin-film transistors with polymeric gate dielectric. *Org. Electron.* **2004**, *5*, 67–71. [CrossRef]
28. Liu, Z.; Yin, Z.; Chen, S.C.; Dai, S.; Huang, J.; Zheng, Q. Binary polymer composite dielectrics for flexible low-voltage organic field effect transistors. *Org. Electron.* **2018**, *53*, 205–212. [CrossRef]
29. Noh, Y.Y.; Sirringhaus, H. Ultra-thin polymer gate dielectrics for top-gate polymer field-effect transistors. *Org. Electron.* **2009**, *10*, 174–180. [CrossRef]
30. Cheng, X.; Caironi, M.; Noh, Y.Y.; Wang, J.; Newman, C.; Yan, H.; Facchetti, A.; Sirringhaus, H. Air Stable Cross-Linked Cytop Ultrathin Gate Dielectric for High Yield Low-Voltage Top-Gate Organic Field-Effect Transistors. *Chem. Mater.* **2010**, *22*, 1559–1566. [CrossRef]
31. Ha, Y.G.; Jeong, S.H.; Wu, J.S.; Kim, M.G.; Dravid, V.P.; Facchetti, A.; Marks, T.J. Flexible Low-Voltage Organic Thin-Film Transistors Enabled by Low-Temperature, Ambient Solution-Processable Inorganic/Organic Hybrid Gate Dielectrics. *J. Am. Chem. Soc.* **2010**, *132*, 17426–17434. [CrossRef] [PubMed]
32. Kim, S.H.; Hong, K.P.; Jang, M.; Jang, J.Y.; Anthony, J.E.; Yang, H.; Park, C.E. Photo-Curable Polymer Blend Dielectrics for Advancing Organic Field-Effect Transistor Applications. *Adv. Mater.* **2010**, *22*, 4809–4813. [CrossRef] [PubMed]
33. Vidor, F.F.; Meyers, T.; Hilleringmann, U. Inverter Circuits Using ZnO Nanoparticle Based Thin-Film Transistors for Flexible Electronic Applications. *Nanomaterials* **2016**, *6*, 154. [CrossRef] [PubMed]
34. Li, S.; Zhang, Q. The dielectric properties of low temperature thermally cross-linked polystyrene and poly(methyl methacrylate) thin films. *RSC Adv.* **2015**, *5*, 28980. [CrossRef]
35. Li, S.; Feng, L.; Zhao, J.; Guo, X.; Zhang, Q. Low temperature cross-linked, high performance polymer gate dielectrics for solution-processed organic field-effect transistors. *Polym. Chem.* **2015**, *6*, 5884–5890. [CrossRef]

36. Xu, W.; Guo, C.; Rhee, S.W. High performance organic field-effect transistors using cyanoethyl pullulan (CEP) high-k polymer cross-linked with trimethylolpropane triglycidyl ether (TTE) at low temperatures. *J. Mater. Chem. C* **2013**, *1*, 3955–3960. [CrossRef]
37. Yoon, M.H.; Yan, H.; Facchetti, A.; Marks, T.J. Low-Voltage Organic Field-Effect Transistors and Inverters Enabled by Ultrathin Cross-Linked Polymers as Gate Dielectrics. *J. Am. Chem. Soc.* **2005**, *127*, 10388–10395. [CrossRef] [PubMed]
38. Li, S.; Tang, W.; Zhang, W.; Guo, X.; Zhang, Q. Cross-linked Polymer-Blend Gate Dielectrics through Thermal Click Chemistry. *Chem. Eur. J.* **2015**, *21*, 17762–17768. [CrossRef] [PubMed]
39. Lai, S.; Cosseddu, P.; Zucca, A.; Loi, A.; Bonfiglio, A. Combining inkjet printing and chemical vapor deposition for fabricating low voltage, organic field-effect transistors on flexible substrates. *Thin Solid Films* **2017**, *631*, 124–131. [CrossRef]
40. Leppaniemi, J.; Huttunen, O.H.; Majumdar, H.; Alastalo, A. Flexography-Printed In$_2$O$_3$ Semiconductor Layers for High Mobility Thin-Film Transistors on Flexible Plastic Substrate. *Adv. Mater.* **2015**, *27*, 7168–7175. [CrossRef] [PubMed]
41. Vidor, F.F.; Wirth, G.I.; Hilleringmann, U. Low temperature fabrication of a ZnO nanoparticle thin-film transistor suitable for flexible electronics. *Microelectron. Reliab.* **2014**, *54*, 2760–2765. [CrossRef]
42. Jung, S.Y.; Albariqi, M.; Gruntz, G.; Al-Hathal, T.; Peinado, A.; Garcia-Caurel, E.; Nicolas, Y.; Toupance, T.; Bonnassieux, Y.; Horowitz, G. A TIPS-TPDO-tetraCN-Based n-Type Organic Field-Effect Transistor with a Cross-linked PMMA Polymer Gate Dielectric. *ACS Appl. Mater. Interfaces* **2016**, *8*, 14701–14708. [CrossRef] [PubMed]
43. Ha, Y.G.; Emery, J.D.; Bedzyk, M.J.; Usta, H.; Facchetti, A.; Marks, T.J. Solution-Deposited Organic-Inorganic Hybrid Multilayer Gate Dielectrics. Design, Synthesis, Microstructures, and Electrical Properties with Thin-Film Transistors. *J. Am. Chem. Soc.* **2011**, *133*, 10239–10250. [CrossRef] [PubMed]
44. Wang, B.; Carlo, G.D.; Turrisi, R.; Zeng, L.; Stallings, K.; Huang, W.; Bedzyk, M.L.; Beverina, L.; Marks, T.J.; Facchetti, A. The Dipole Moment Inversion Effects in Self-Assembled Nanodielectrics for Organic Transistors. *Chem. Mater.* **2017**, *29*, 9974–9980. [CrossRef]
45. Kang, W.G.; An, G.I.; Kim, M.J.; Lee, W.H.; Lee, D.Y.; Kim, H.J.; Cho, J.H. Ladder-Type Silsesquioxane Copolymer Gate Dielectrics for High-Performance Organic Transistors and Inverters. *J. Phys. Chem. C* **2016**, *120*, 3501–3508. [CrossRef]
46. Kim, Y.T.; Roh, J.K.; Kim, J.H.; Kang, C.M.; Kang, I.N.; Jung, B.J.; Lee, C.H.; Hwang, D.H. Photocurable propyl-cinnamate-functionalized polyhedral oligomeric silsesquioxane as a gate dielectric for organic thin film transistors. *Org. Electron.* **2013**, *14*, 2315–2323. [CrossRef]
47. Lee, D.H.; Jeong, H.D. Solution-Processed Gate Insulator of Ethylene-Bridged Silsesquioxnae for Organic Field-Effect Transistor. *J. Chosun Nat. Sci.* **2010**, *3*, 7–18.
48. Hamada, T.; Nagase, T.; Kobayashi, T.; Matsukawa, K.; Naito, H. Effective control of surface property on poly(silsesquioxane) films by chemical modification. *Thin Solid Films* **2008**, *517*, 1335–1339. [CrossRef]
49. Choi, S.S.; Lee, A.S.; Lee, H.S.; Jeon, H.Y.; Baek, K.Y.; Choi, D.H.; Hwang, S.S. Synthesis and Characterization of UV-Curable Ladder-Like Polysilsesquioxane. *J. Polym. Sci. Part A Polym. Chem.* **2011**, *49*, 5012–5018. [CrossRef]
50. Park, E.S.; Ro, H.W.; Nguyen, C.V.; Jaffe, R.L.; Yoon, D.Y. Infrared Spectroscopy Study of Microstructures of Poly-(silsesquioxane)s. *Chem. Mater.* **2008**, *20*, 1548–1554. [CrossRef]
51. Ni, Y.; Zheng, S. A Novel Photocrosslinkable Polyhedral Oligomeric Silsesquioxane and Its Nanocomposites with Poly(vinyl cinnamate). *Chem. Mater.* **2004**, *16*, 5141–5148. [CrossRef]

© 2018 by the authors. Licensee MDPI, Basel, Switzerland. This article is an open access article distributed under the terms and conditions of the Creative Commons Attribution (CC BY) license (http://creativecommons.org/licenses/by/4.0/).

Article

Omnidirectional SiO₂ AR Coatings

Sadaf Bashir Khan [1], Hui Wu [1] and Zhengjun Zhang [2,*]

[1] The State Key Laboratory for New Ceramics & Fine Processing, School of Materials Science & Engineering, Tsinghua University, Beijing 100084, China; sadafbashirkhan@yahoo.com (S.B.K.); huiwu@tsinghua.edu.cn (H.W.)
[2] Key Laboratory of Advanced Materials (MOE), School of Materials Science & Engineering, Tsinghua University, Beijing 100084, China
* Correspondence: zjzhang@tsinghua.edu.cn; Tel.: +86-10-6279-7033

Received: 18 April 2018; Accepted: 26 May 2018; Published: 1 June 2018

Abstract: It is of great importance to develop antireflective (AR) coatings and techniques because improved optical performance has been progressively prerequisite for wide-ranging applications such as flat panel displays, optoelectronic devices or solar cells. Natural, surroundings inspire researchers considerably to impersonate in order to provoke analogous characteristics via artificial approaches, which provide the opportunity for emerging techniques and development in material engineering. Herein, SiO_2 antireflective (AR) coatings comprised of two layers were fabricated using a physical vapour deposition method via glancing angle. The, top layer fabricated at an oblique angle of 80° and the bottom layer close to the substrate was deposited at a deposition angle of 0°. The, experimental outcomes demonstrate that there is a slight influence on the refractive index of thin films by changing the morphology of nanostructures keeping deposition angles the same. The, top layer shows a periodic arrangement of SiO_2 nanostructures while the bottom stratum represents a SiO_2 compact dense layer. The, assembled bilayer SiO_2 AR coating retains omnidirectional AR efficiency and tunability at a preferred wavelength range displaying <1% reflectance. Moreover, the fabricated omnidirectional SiO_2 AR coatings have thermal stability up to 300 °C. These, SiO_2 AR coatings also possess negative temperature resistivity to withstand different cold storage conditions. Hence, the flexible and environmental adaptive SiO_2 AR coating offers an intriguing route for imminent research in optics.

Keywords: refractive index; deposition angle; wavelength; antireflective; omnidirectional; nanostructures; thermal stability

---

1. Introduction

Different optoelectronics instruments such as eyeglasses, cathode ray tubes, display panels, solar cells covers, or windows require anti-reflective coatings (ARCs) [1]. A light reflection from optical boundaries in a coating stack is an outcome of dissimilarities in the refractive index profile in a coating [2,3]. In, optoelectronic instruments including solar thermal cells, monitors, liquid crystal displays, telescopes, and shielding windows in greenhouses, AR coatings have a great role in diminishing undesirable light reflectance to enhance the overall working potential of transmissive optical features [4–8]. AR, coatings were used to suppress Fresnel reflections between the surrounding media (air) and AR coated substrate interface [9]. Augustin-Jean Fresnel [10] introduces the concept of reflectance loss when light strikes the interfaces having different media. The, light impinges on glass having refractive index of 1.52 shows Fresnel reflection at the air-glass interface with 4.2% reflectance loss at 550 nm. An, AR coating significantly reduces the resultant losses (4% per interface) by using index-matching materials.

Earlier, numerous groups formulated AR coatings comprised of organic, inorganic or composite materials [11,12]. Diverse, composites or amalgam nanostructures were assembled to produce single-layer, bilayer or multilayer AR coatings [13,14]. A single layer AR coating reduces reflectance only at a specified preferred wavelength (quarter-wave (QW)) [15]. A double layer or multilayer AR coatings can diminish reflection in a broader region (more than one wavelength of interest), but the preference of the material selection is more limited, due to refractive index restraints [16,17]. Multifunctional, AR coatings with superhydrophobicity, transparency, and thermal stability possessing an ultralow refractive index have been studied by many research groups [18]. These, AR coatings have promising potential applications in industrial and commercial scale in electronic or optical devices. However, the poor durability of anti-reflective coatings is a great obstacle hindering their applications [11,19]. A bilayer or trilayer composite AR coating of SiO_2-TiO_2 shows 99% transmittance at the wavelength of 351 and of 1053 nm, and the tri-layer AR film shows nearly 100% transmittance at the wavelengths of 527 and 1053 nm suitable for laser applications [20]. Plastic, substrates such as PMMA shows 99% transmittance in a visible region when a silica layer is coated on both sides of substrates [21]. However, the key disadvantage and limitation of multilayer AR thin film occurs due to appropriate refractive index selection, property incompatibility at the layer edges, and the applicable arrangement of a discrete layer in a multilayer coating stack. These, issues make the multilayer composite AR coatings production development complex and bind it for applying them commercially due to expensive costs. Moreover, in a composite multilayer ARC, the increase in a number of layers generates intrinsic problems including the refractive index constancy, precise thickness control of discrete layer throughout the coating stack, and impurity fortification from the adjoining environments. It, is easy to fabricate ARCs comprised of a single material to avoid the concerns of choosing the pertinent materials with matching appropriate steady properties.

The most interesting approach for broadband AR coating encompasses the creation of gradient configurations on wavelength scale such as nanopillar fabrication mimicking moth eye. Graded, index AR coatings enhance efficiency at a wider spectral region. Kennedy, and Brett study the humidity impact on refractive index of graded-index SiO_2 AR films fabricated via glancing angle deposition on glass [22], composite AR coatings of quarter wave double-layer (TiO_2-SiO_2, 94.4% transmittance), (ZrO_2-SiO_2, 94.3% transmittance) or groove surfaces of bilayer or trilayer AR coatings used in solar cells to enhance solar cell efficiency [23,24]. Zhang [25] reported a simple way to modify refractive index (1.10–1.45) of SiO_2 established on SiO_2 hollow nanospheres based on hybridization with acid-catalyze on a low iron glass substrate. The, hybridization process impacts and influences the refractive index, thin-film thickness and roughness. They, fabricate single-layered ARCs showing 99% transmittance at a single wavelength of 600 nm and the three-layered ARCs (300–800 nm) showing 97.29% transmittance. Similarly, mesoporous SiO_2 nanospheres show AR efficiency and reduce glass reflectance from 8% to <2% [26]. SiO_2 AR coatings for UV laser applications [27] and ammonia-SiO_2 composite monolayer AR coating by the sol-gel method enhance laser-disk pumping efficiency [28]. Wang, fabricates antibacterial composite AR coatings comprised of mesoporous SiO_2 with Ag nanoparticles. The, optimal thin-film-coated glass substrate exhibits an average transmittance of 97.1%, in the range of 400–800 nm [29].

Diverse top down or bottom up approaches were used to fabricate AR coatings including sputtering process [30], liquid phase deposition [31,32], interference or electron beam lithography [33] to replicate structures. However, the majority of these approaches are characteristically challenged due to intricate complex processes with various production stages. In, the sol-gel fabrication process, extensive sol ageing times causes trouble. In, addition, organic constituent compatibility, its effects on the condensation or hydrolysis process, worsening of organic functionality due to high-temperature exposure, contamination or dust particle influence limit its practical approach and inflict complex challenges. The, main shortcoming of the sol-gel fabrication method is the narrow choice of supporting substrates. Previously, most of SiO_2 AR coating was assembled on K9 or glass substrates via sol-gel technique [34]. The, pore densification step requires high sintering temperature (200–500 °C), which restricts the fabrication to few substrates [35]. Regarding, this, the physical vapour deposition (PVD) process is most applicable for production of ARCs as direct deposition has a minimal probability of foreign contamination in the fabricated films. The, fabricated films hold good compositional uniformity over film thickness in nanorange, displaying reliability and steadiness in the refractive index profile [22,36].

The purpose of the current work is to fabricate omnidirectional AR coatings and also to study the influence of morphology and its impact on thin films' refractive index under identical fabrication parameters. Here, we demonstrate a simple method to fabricate SiO_2 AR coating using the PVD technique. This, methodology permits efficient fine-tuning of the porosity of the individual stratum in a bilayer coating stack to reduce reflectance in a broader wavelength region. The, optical performance of the fabricated ARCs was calculated under different environmental conditions. The, experimental result shows that there is a very slight influence on the refractive index by changing the morphology keeping the deposition angles same. The, bottom layer in a bilayer coating stack is dense and similar in all fabricated SiO_2 AR coatings. The, morphology of the top porous layer consists of nanozigzag, nanohelix, C-shape and slating nanostructures. All, of the bilayer coatings show good omnidirectional AR efficiency (<1%) in the visible spectra, but the one having zigzag morphology shows the best among all due to higher porosity and a gradual variation in the refractive index profile from top to bottom, which enhances AR efficiency overall.

The novelty of the present work on SiO_2 AR coatings lies in the fabrication mechanism and engineering nanostructure. The, current coating design is reliable, easy to formulate on a large scale, and applicable on different transparent or semitransparent supporting substrates. One, can simply lessen the reflectance of the preferred substrate only by modifying the distinct stratum thickness and choosing the applicable refractive index in a coating stack through modelling and simulation of refractive indices suitable for minimizing reflectance. One, of the significant characteristics of the strategy is the non-existence of property incompatibility impact at the interfaces between stratum and least contamination influence as the binary layers encompass a single material.

2. Materials and Methods

2.1. Materials

Borosilicate crown glass (BK7) having dimensions 15 mm × 15 mm and silicon wafers were used as supporting substrates to deposit SiO_2 thin films. Before, the fabrication process, the supporting substrates BK7 were washed in an ultrasonic bath with ethanol and acetone for half an hour to remove any kind of contamination. Afterwards, the BK7 was dipped in deionized water for five to ten minutes and dried in an atmospheric environment.

2.2. Fabrication Method

Glancing angle deposition technique is used to fabricate bilayer SiO_2 AR coatings under a high vacuum using an electron beam evaporation machine at room temperature. A quartz-crystal microbalance (QCM) positioned close to the substrate is used to evaluate the thickness growth of the individual thin layer. During, experiments, the fabrication parameters i.e., the deposition rate (5 A°/s) and base pressure (3×10^{-4} Pa), were kept similar. According, to the experimental needs, we can change and incline the substrate at the different tilted position. The, substrate is positioned at a specified distance of 10 cm away from the target (SiO_2 99.9% high purity). The, SiO_2 AR coatings assembled on silicon wafers were used for morphological study and refractive index study while the AR coatings fabricated on BK7 transparent substrates were used to evaluate AR characteristics, omnidirectionality, thermal stability and negative temperature influence.

2.3. Characterization

Field Emission Scanning Electron Microscopy SEM, JEOL-7001F (operational voltage; 15 kV) (JEOL, Tokyo, Japan) is used to analyse the cross-sectional nanostructure and top morphology of as-deposited bilayer SiO_2 AR coating. Computer, software is used for simulating SiO_2 AR coatings by adjusting distinct stratum width in the coating structure. In, order to determine the refractive index of SiO_2

nanofilms at a wavelength of 550 nm, a WVASE32 spectroscopic ellipsometer instrument (JA Woollam Co., Inc., Lincoln, NE, USA) is used. AR, performance of bilayer SiO_2 at the normal light incidence and oblique omnidirectional angles were supported by means of an ARM (angle-resolved microscope) R1 series Ideaoptics (HL2000 Pro) instrument (IDEAOPTICS Instrument Co., Ltd., Shanghai, China). AR, measurements were performed at five different locations on the sample surface to determine the homogeneity of thin films. The, average result of five readings was reported here for accuracy. The, surface wettability of SiO_2 AR coatings was determined by using a drop shape analyzer instrument (DSA 100, Krüss GmbH, Hamburg, Germany). A needle having a diameter of 0.5 mm and length of 38 mm is used to measure the water contact angle using a droplet volume of 5 μL.

2.4. Optical Simulation for Designing AR Coating

Thin Film Calculator software (TFcalc) (3.5, Software Spectra, Inc., Portland, OR, USA), a computer controlled programming method is used for simulation purposes to design AR coatings. This, software helps in designing different kinds of AR coatings including broadband, multilayers, narrow band or single wavelength AR coatings in accordance with preferred specifications. During, simulation, the input parameters comprise the number of layers in coating stack, η of the individual stratum, layer thickness, light incident angle, substrate η, one-sided substrate reflectance consideration and preferred wavelength region. The, output parameter consists of reflectance curves of bilayer AR coating. Preliminary, studies illustrate that the two layers selected for assembling AR coating were a porous layer having η = 1.17–1.19 adjoining air media and a compact layer having η = 1.46 close to the substrate. The, two layers are arranged in a manner to create a gradual increment in a refractive index from surrounding air media towards the substrate bottom. This, layer arrangement helps in reducing the reflectance of substrate material increasing AR efficiency. Layer, thickness and the refractive index of the individual layer in a coating stack are of vital importance to enhance AR characteristics. In, the current work, firstly TFCalc software is used to optimize the thickness of each individual layer and the arrangement of the specified layer to generate low reflectance in the visible desired region through simulation. The, bilayer AR coating is fabricated by using the simulation parameters. The, assemble bilayer AR coating shows nearly the same AR performance in accordance with the simulated design AR coating.

3. Results and Discussion

3.1. Morphological Study of SiO_2 Single Layer Nanostructures

Firstly, we deposit SiO_2 nanofilms at an oblique angle of 80° and try to change the morphology of a single silica layer by controlling the fabrication parameters including the substrate rotational speed by keeping the deposition angle and base pressure similar. Herein, we try to study the morphological influence on the refractive index and porosity of thin films.

3.1.1. Nanozigzag, Nanohelix, Slanting and C-Shape SiO_2 Nanofilms Fabrication

SiO_2 thin films having different morphologies were deposited on silicon wafers in a high vacuum e-beam evaporation system with a base pressure of 3×10^{-4} Pa by using a glancing angle deposition technique. The, oblique incident was adjusted at, θ = 80° to fabricate nanozigzag, nanohelix, slanting and C-shape SiO_2 single layer thin films. Different, nanofilms were fabricated in order to study the influence of morphology on the refractive index of thin films. The, experimental result shows that there is a slight increment or decrement in the refractive index of thin films by changing the morphology. The, growth mechanics of SiO_2 nanostructures is mentioned below.

3.1.2. SiO_2 Nanostructures Growth Mechanism

The governing principle that leads towards the nanostructure growth is based on a self-shadowing effect and adatom diffusion. The, nanostructures column development proceeds because atomic shadowing generates zones. Due, to these zones, the vapour flux cannot directly reach the substrate. The, adatom mobility is very low for surface diffusion to fill the cavities and voids. When, the vapour flux reaches at a glancing angle ($\alpha > 70°$), the atomic shadowing influence is considerably boosted. This, leads to a formation of porous columnar nanostructure morphology of isolated grains directed towards the vapour source. Thus, columnar nanostructure was generated when the substrate is tilted at $\alpha > 70°$ during deposition. The, nanostructures do not develop and advance in a parallel direction as the direction of the incident vapour flux, but they always develop and grow towards the vertical direction.

According to experimental requirements, the growing nanostructures morphology can be modified easily by monitoring the substrate rotational condition, stationary position, deposition angle, or the vapour flux arrival during the fabrication process. We, also fabricate different morphologies of SiO_2 nanostructures such as nanozigzag, nanohelix, slanting and C-shape SiO_2 nanofilms by controlling and adjusting the substrate position as shown in Figure 1. Figure 1a represents the SiO_2 nanozigzag film. The, zigzag nanostructure can be easily established by instigating the vapour flux arrival direction to rotate the substrate holder by 180° repeatedly in clockwise and anticlockwise directions while the deposition angle was kept constant at 80° throughout the deposition process. Similarly, C-Shape nanofilms and nanohelics can be fabricated by rotating substrates at a constant rate comparative to the deposition rate. Thus, a columnar C-shape (0.06 Rev/min) or nanohelics (0.12 Rev/min) formation takes place comprised of constant pitches depending on the rotational speed of the substrate with respect to revolution per time.

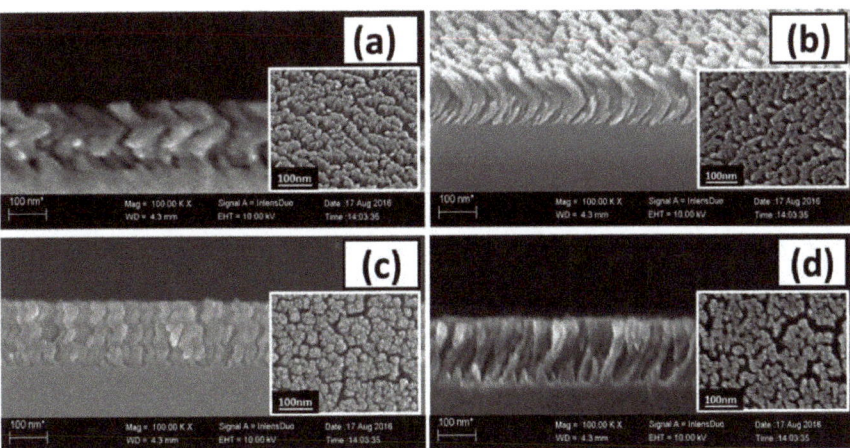

Figure 1. Cross-sectional images of SiO_2 nanostructures: (**a**) nanozigzag, (**b**) C-Shape, (**c**) nanohelix, and (**d**) slanting nanorods. The, inset in each figure represents the top view morphology of thin films.

Slanting nanorods as shown in Figure 1d is fabricated by a slanting substrate holder with respect to the target normal direction by an angle of 80° keeping the substrate holder stationary, and thickness of the developing film is controlled by a deposition rate measured via QCM. The, QCM determines variation in mass flux per unit area by computing the frequency change. When, target materials having high melting points are deposited at room temperature under high vacuum in a tilt substrate position, then the vapour flux arrives at a normal substrate, which results in growing columnar nanostructures. SEM, images observably indicate that deposition angle plays a key role in monitoring the nanostructure morphology of thin films due to limited atom mobility and self-shadowing effect. Cross-sectional SEM images show that, in all the fabricated SiO_2 nanostructures, the inter-column spaces increase, due to which extremely oriented, nanocolumn development proceeds. The, nanocolumns in spite of any shape show good separation, consistency, uniformity and homogeneity with periodical void spaces between growing nanostructures as shown in Figure 1.

3.2. Refractive Index Analysis

The refractive index of SiO$_2$ nanostructures was determined by using a WVASE32 spectroscopic ellipsometer (JA Woollam Co., Inc., Lincoln, NE, USA) in the wavelength range of 300–900 nm. The, η of SiO$_2$ nanostructures were calculated by using the Cauchy dispersion model, as SiO$_2$ is transparent and dielectric. The, experimental results show that there is a slight variation in the refractive index of thin films at 550 nm wavelength, as shown in Figure 2.

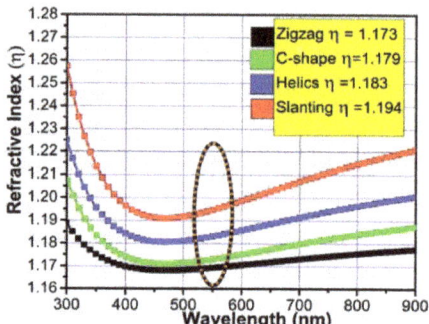

Figure 2. Refractive index (η) of SiO$_2$ nanozigzag, nanohelix, slanting and C-shape nanofilms measured as a function of wavelength at 550 nm.

The nanozigzag and C-shape SiO$_2$ nanostructures nearly show the same η = 1.17. However, SiO$_2$ nanohelix structures show the η = 1.18. Earlier, in our previous experiments, we deposited standing nanorods that almost show the same refractive index of 1.18 [24]. Slanting, SiO$_2$ nanorods show the highest refractive index of 1.19 among all fabricated nanostructures at a deposition angle of 80°. The, experimental results prove that overall there is a very slight influence on the refractive index of thin films by changing the morphology of nanostructures keeping the deposition angle the same. The, η value remains in between 1.17–1.19 at angle 80° by changing morphology. In, thin films, the main cause for the decrement in the effective refractive index is due to an upsurge in porosity, which influences the packing density and mass fraction. Figure 2 indicates that nanozigzag SiO$_2$ film shows higher porosity in comparison with other nanostructures. Overall, all of the SiO$_2$ nanofilms are very permeable, spongy and porous comprised of mass flux and void air spaces.

In order to prove and validate our findings of the nanostructure influence on the refractive index, we use the maximum glancing angle 88° and fabricate nanostructures with varying morphologies including standing [24], and C-shape nanofilms as shown in Figure 3. The, experimental results show that, at such high oblique angle, there are minor differences in refractive index still existing at the nanolevel. The, SEM top and cross-sectional images clearly indicate that nanofilms are highly oriented, well-spaced, homogenous, and separated. Figure 3a,b shows the standing nanorods, and C-shape nanofilms fabricated at a glancing angle of 88° with fast substrate rotational and slow rotation. Figure 3c shows the refractive index of SiO$_2$ standing, and C-shape measured as a function of wavelength at 550 nm. The, refractive index analysis shows that there is a very slight negligible influence of refractive index at a glancing angle of 88°. Overall, morphology did not impact the refractive index at 88° and it lies in between 1.07–1.08.

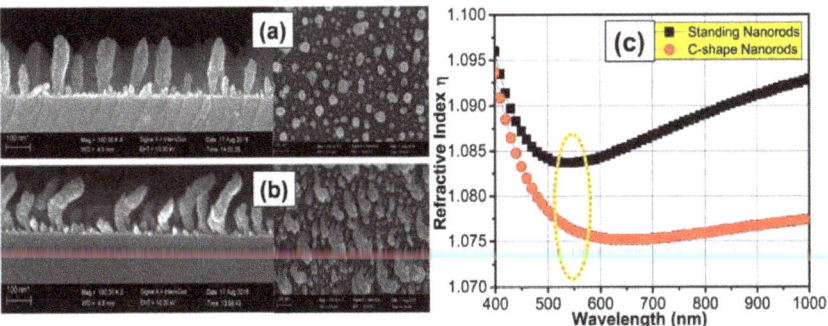

Figure 3. Shows the (**a**) standing nanorods (reproduced with permission from [34]; © 2018 WILEY-VCH Verlag GmbH & Co. KGaA), and (**b**) C-shape nanofilms fabricated at glancing angle of 88°; (**c**) the refractive index of SiO$_2$ standing and C-shape nanofilm calculated as a function of wavelength at 550 nm.

3.3. Fabrication of SiO$_2$ AR Coatings

In fabricating AR coatings, optical thickness and η play an important role in reducing the reflectance of the desired substrate. In, our experiments, we deposit bilayer SiO$_2$ AR coatings comprised of a dense layer and a porous layer on top of the dense layer to reduce the reflectance of BK7 glass substrate less than 1 percent in the visible region. Firstly, the dense film is deposited on the forward-facing side at deposition angle 0° having a refractive index of 1.46 at 550 nm wavelength [34]. On, the top of the dense film, a porous vertical nanostructure having lower η than the bottom layer was deposited at an oblique angle (α) of 80°.

Both layers were deposited in a single step just by changing the tilting position of the substrate to avoid any foreign contamination or dust influence. The, oblique angle was preferred to induce porosity as well as to generate a steady deviation in η of AR coating stack from surrounding air towards the substrate bottom. Figure 4 clearly represents the cross-sectional SEM image of bilayer SiO$_2$ AR coatings having different morphologies on BK7 substrate. The, SEM image clearly shows the interfaces between the dense layer and the top porous layer having different morphologies comprised of nanozigzag, C-shape, nanohelix, and slanting nanostructure. In, all cases, the bottom layer is deposited at angle 0° and the top layer is deposited at angle 80°. The, edges of two layers in the SiO$_2$ AR stack and the supporting substrate is clearly evident, indicating the consistency and regularity within the individual layer throughout the thickness of nanostructure representing stability and consistency in the refractive index profile.

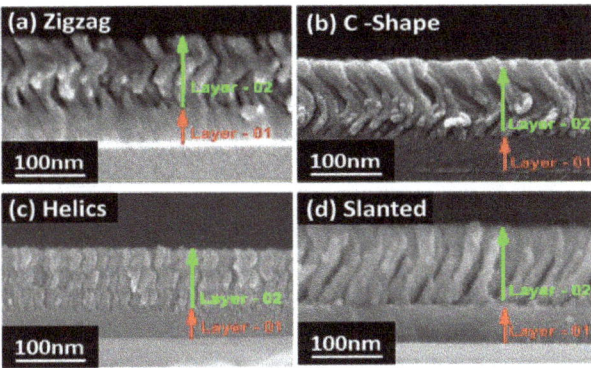

Figure 4. Bilayer SiO$_2$ AR coatings having different morphologies on BK7 substrate: (**a**) nanozigzag, (**b**) C-shape, (**c**) nanohelix, and (**d**) slanting nanostructure.

3.4. AR Efficiency of SiO₂ AR Coatings

The angle-resolved microscope (ARM) R1 series Ideaoptics is used to determine the reflectance properties of bilayer SiO$_2$ AR coatings at the normal light incidence and omnidirectional angles. BK7 substrate reflects 4.26%–5% in the wavelength range between 500–900 nm. Fabricating, bilayer SiO$_2$ AR coating reduces the reflectance of BK7 to <1% reflectance in the wavelength range 500–900 nm as shown in Figure 5. The, measured experimental reflectance (dE) outcomes of fabricated bilayer SiO$_2$ AR coating are quite analogous to each other displaying <1% reflectance as shown in Figure 5 in the visible region.

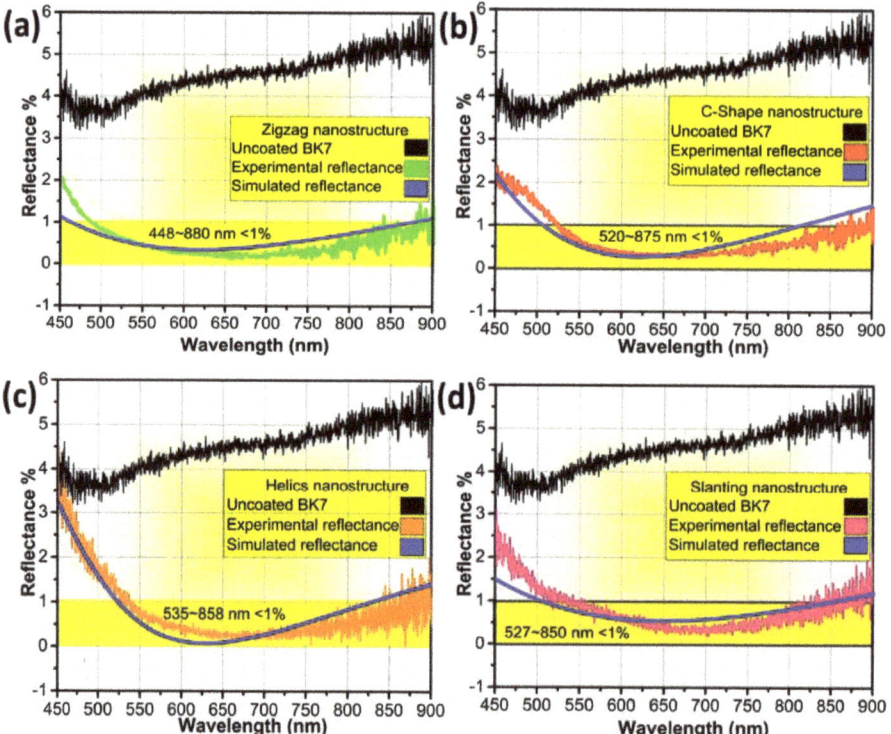

Figure 5. Bilayer SiO$_2$ AR coating efficiency on BK7 substrate: (**a**) nanozigzag, (**b**) C-shape, (**c**) nanohelix, and (**d**) slanting nanostructures. Blue, curves represent the simulated reflectance of design coating. The, black curve represents reflectance from the uncoated substrate. Coloured, curves represent the experimentally measured reflectance results of fabricated bilayer SiO$_2$ AR coatings.

The fabrication parameters i.e., the deposition angle and individual layer thickness in bilayer SiO$_2$ AR coating stack is mentioned in Table 1. The, experimental result shows that AR efficiency of nanozigzag is better than the other SiO$_2$ AR coatings having different morphologies. Nanozigzag, bilayer SiO$_2$ AR coating shows less <1% reflectance in the whole visible region 470–850 nm at normal incidence angle. C-shape and nanohelix bilayer SiO$_2$ AR coatings show nearly identical AR efficiency (<1%) in the wavelength range of 520–900 nm. Slanting, SiO$_2$ AR coatings shows a slight decrement in efficiency (540–850 nm) in comparison with other SiO$_2$ AR coatings. The, reason is due to a slight variation in refractive index of top layers in comparison with other coatings throughout the wavelength range (300–900 nm).

Here, the bilayer SiO$_2$ AR films were fabricated according to simulation parameters. The, simulated AR curves are similar to experimental reflectance curves of as-deposited SiO$_2$ AR films, as shown in Figure 5a–d. The, thickness of each individual layer in bilayer coating stack information is taken by simulated design AR film as mentioned in Table 1. The, experimental fabricated thickness of bilayer SiO$_2$ AR films is identical to the thickness of the individual layer of a simulated design AR coating stack. The, simulated thickness is represented by dS and experimentally fabricated AR film thickness is abbreviated as dE, as stated in Table 1. The, experimental results show that, by optimizing the thickness of the individual layer in a coating stack and selecting the appropriate refractive index, one can design AR coatings according to optical instrument requirements at the desired wavelength region.

Table 1. Bilayer SiO$_2$ AR coating fabrication parameters.

Morphology	Deposition Angle (α)		Refractive Index (η)		Thickness (nm) dS		Thickness (nm) dE	
	Layer-1	Layer-2	Layer-1	Layer-2	Layer-1	Layer-2	Layer-1	Layer-2
Nanozigzag	0	80	1.45	1.173	65	130	70	130
C-shape	0	80	1.45	1.179	75	135	78	141
Nanohelix	0	80	1.45	1.18	60	142	56	147
Slanting	0	80	1.45	1.19	64	130	56	125

3.5. Omnidirectional AR Efficiency of SiO$_2$ AR Coatings

Practically in different optoelectronic appliances, AR coating encompassing wideband reflectance over a wide light incident angles (AOI) is mandatory and prerequisite. Here, we also determine the AR efficiency of our fabricated coatings at a different angle of light incidence via full angle reflection (FAR) operational mode. The, omnidirectional AR capacity of bilayer SiO$_2$ AR coating having different morphologies is shown in Figure 6.

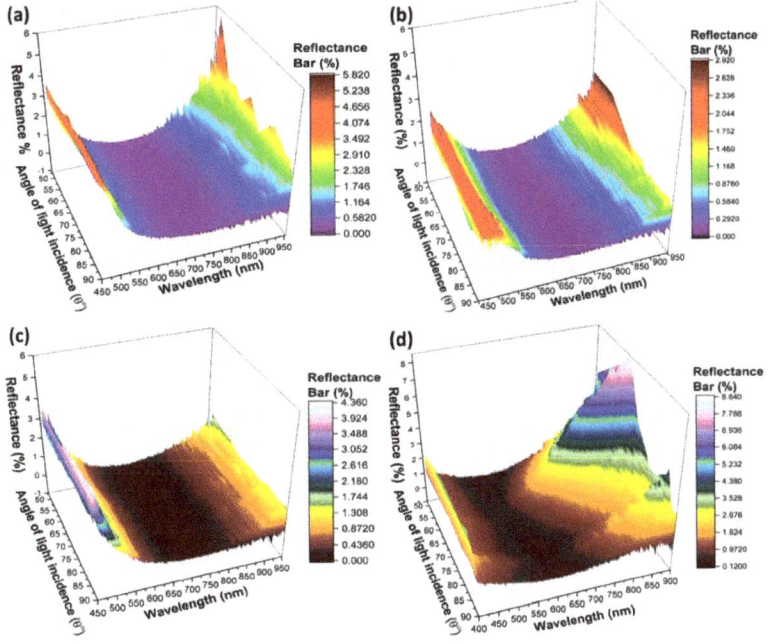

Figure 6. Omnidirectional AR efficiency of bilayer SiO$_2$ AR coating on BK7 substrate at a different angle of light incidence: (**a**) nanozigzag, (**b**) C-shape, (**c**) nanohelix, and (**d**) slanting nanostructure.

The experimental measurement demonstrates that, in the visible wavelength range, bilayer SiO_2 AR coating having zigzag morphology preserves its AR proficiency uniformly at oblique angles. The, zigzag SiO_2 AR coating reflecting 0.01%–0.58% light in the wavelength range of 530–850 nm at oblique angles varies from normal incidence 90° up to 50° with 5° change. Similarly, the bilayer SiO_2 AR coating having different morphologies Figure 6b C-shape, Figure 6c nanohelix and Figure 6d slanting nanostructure also show omnidirectional AR efficiency at a different angle of light incidence. In, C-shape SiO_2 AR coatings, the films show <1% reflectance in the entire visible spectra. In, the wavelength range of 650–750 nm, C-shape SiO_2 AR coatings show <0.29% light, while, at 550–650 nm and 750–850 nm, it shows <0.58% showing good omnidirectional characteristics.

A similar kind of omnidirectional efficiency is observed in helical SiO_2 AR coatings displaying <0.48% light in the wavelength range of 550–850 nm wavelength. Thus, the experimental result shows that fabricated AR coating is omnidirectional. In, case of slanting SiO_2 AR coatings (Figure 6d), the coating shows <0.5% reflectance in the wavelength range between 550–700 nm at different angles of light incidence. However, at oblique angles of 60° and 55°, the AR efficiency declines towards higher wavelength regions (650–750 nm) showing nearly 1%–2% reflectance. Beyond, this wavelength (i.e., 750–850 nm), the AR film reflects nearly 2%–3% light. Overall, the performance of AR coatings having different morphologies is stable and consistent in visible spectra showing an incident-angle-insensitive antireflective omnidirectional characteristic.

3.6. Negative Temperature Stability

Experiments were also carried at a negative temperature to investigate the AR efficiency of nanozigzag bilayer SiO_2 coating. We, keep the bilayer SiO_2 AR films at 32 °F (0 °C, refrigerator) and 0 °F (−18 °C) for 24 h in order to study the negative temperature impact on the AR efficiency of as-deposited bilayer SiO_2 coating. The, experimental result displays that the AR film retains its AR efficiency even when exposed to such low temperatures. The, as-deposited and negative exposed nanozigzag bilayer SiO_2 coating shows similar AR efficiency when measured at normal incidence angle. There, is a very negligible variation in AR efficiency (towards lower wavelength region 500 nm) of as-deposited SiO_2 AR coating and cold storage SiO_2 AR coating after exposure to 24 h at 0 °C, and −18 °C, as shown in Figure 7. The, inset in Figure 7 shows SEM images of as-deposited and cold storage nanozigzag bilayer SiO_2 AR coatings, showing that there is no change in morphology taking place. Due, to this, there is consistency and no worsening occurs in AR performance of bilayer SiO_2 thin films.

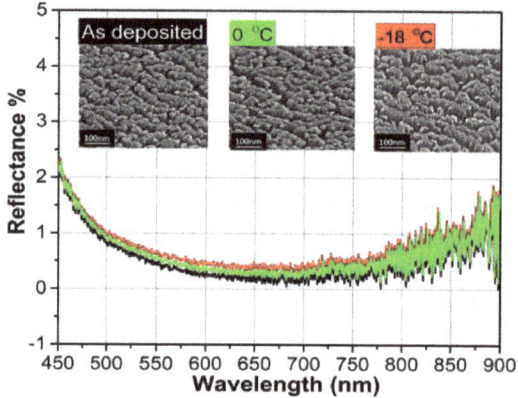

Figure 7. Experimentally calculated AR efficiency of as-deposited nanozigzag SiO_2 AR coating and cold storage SiO_2 coating at 0 °C, and −18 °C, The inset represents the top morphology of as-deposited AR and negative temperature exposed SiO_2 AR coating.

3.7. Annealing Influence on AR Proficiency of SiO$_2$ AR Coatings

Annealing experiments were also performed to examine the AR proficiency of nanozigzag bilayer SiO$_2$ coating at a higher temperature. The, SiO$_2$ AR coating was annealed for an hour at 100, 200, 300 and 350 °C with a ramp speed of 5 °C, as shown in Figure 8. At, 100 °C, the as-deposited SiO$_2$ AR coating and annealed SiO$_2$ AR coating show nearly identical AR efficiency as there is no influence of annealing. At, 200 °C and 300 °C, there is negligible slight declination in AR efficiency in comparison with as-deposited AR coating. However, the SiO$_2$ AR coating shows <1% reflectance in the visible region retaining its AR performance. After, annealing at 350 °C, a clear shift of reflectance minima is observed in comparison with as deposit AR coating. The, AR efficacy of SiO$_2$ AR coating at 350 °C deteriorates due to structural defects, lattice stresses, increased packing density or crystallization [34]. Overall, the SiO$_2$ AR films are thermally stable up to 300 °C without any deterioration in AR efficiency. Beyond, this temperature (350 °C), there is a slight increment in AR efficiency due to induced crystallinity, which influences packing density of thin films in the declining of the AR efficacy [34].

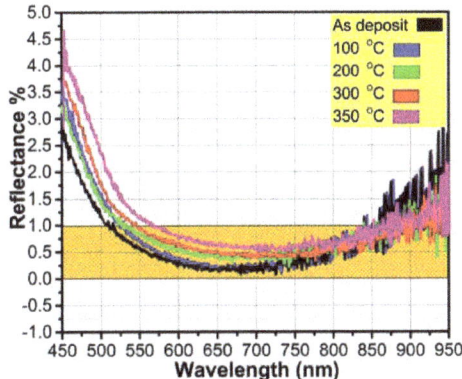

Figure 8. AR efficiency of as-deposited and annealed nanozigzag SiO$_2$ AR coating at 100, 200, 300 and 350 °C.

3.8. Contact Angle Measurement

The wetting behaviour of coatings is a significant phenomenon for its practical usage in real-world applications i.e., TiO$_2$ self-cleaning coatings for photocatalysis [37–40]. The, contact angle (CA) parameter is used to describe the hydrophobicity or hydrophilicity of a surface. In, our experiments, the sessile drop method is used to determine the static contact angle of SiO$_2$ AR coatings having different nanostructures. The, experimental result demonstrates that our fabricated AR coatings are super hydrophilic in nature, showing a water contact angle θ < 12° as shown in Figure 9.

The CA measurements show that SiO$_2$ AR coatings having nanozigzag and C-shape morphology show super hydrophilic behaviour with a CA of nearly 5° and 7°. Nanohelix, and slanting nanostructure also show the same kind of hydrophilic behaviour as shown in Figure 9d,e. The, CA measurements were performed at five different locations on AR coatings and the average result is reported here for accuracy. Figure 9f clearly demonstrates the spreading of water droplets after it drops on SiO$_2$ AR coating. Hydrophilicity, and wettability performance of SiO$_2$ AR coatings depend on nanostructure, chemical composition and surface geometrical dimension. The, hydrophilicity characteristic in our fabricated SiO$_2$ AR coating generates a self-cleaning ability in AR films as the water spreads instantaneously over the surface, rather enduring as droplets. The, hydrophilic behaviour in thin films improves the removal of the dirt particles and makes the surface dry quicker. Thus, SiO$_2$ AR coating offers an intriguing route in different optics applications because of it self-cleaning ability, thermal stability and omnidirectional AR performance.

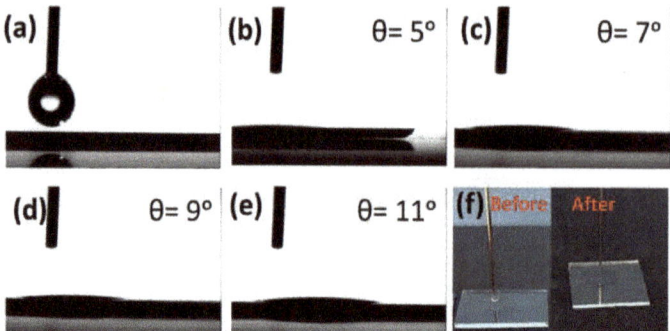

Figure 9. Contact angle measurement of SiO$_2$ AR coating on BK7 substrate (**a**) showing droplet volume of 5 µL use to measure contact angle, (**b**) nanozigzag, (**c**) C-shape, (**d**) nanohelix, (**e**) slanting nanostructure, and (**f**) represents the droplet spreading behavior before and after droplet removal from needle.

4. Conclusions

A PVD method for the fabrication of SiO$_2$ thin films that transforms refractive index from 1.45 to 1.07 has been established by changing the deposition angle. The, omnidirectional optical behaviour of SiO$_2$ AR coating could be easily varied by the rational design of film structure and the choice of the substrate material. It, was experimentally proved that the AR film exhibits a significant decrement in reflectance in comparison with the uncoated supporting substrate. The, outstanding performance of experimentally fabricated coating could be attributed to the design parameters of a bilayer coating on a transparent substrate. The, current coating design advantage is that it is reliable, easy to fabricate, consistent and appropriate for different substrates (transparent or semitransparent). One, can easily reduce the reflectance of the desired substrate only by regulating the discrete stratum thinness and selecting the appropriate refractive index in a coating stack. One, of the noteworthy characteristics of the design is the absence of property incongruity influence at the interface and least contamination impact since the two distinct layers comprise a single structure. Our, experimental result proves that the morphology slightly impacts the refractive index of nanostructures when deposited at the same angle at the nanolevel. The, fabricated bilayer SiO$_2$ AR coatings having different morphologies demonstrate self-cleaning ability, thermal stability and omnidirectional AR efficiency as well as AR constancy when exposed to negative temperature. Engineering, graded-index morphologies at the nanolevel is an effective methodology to engender identical properties such as lotus like superhydrophobicity or being reminiscent of moth eyes. These, kinds of AR coatings are a good choice in optoelectronic devices operational in negative temperature environments.

Author Contributions: Z.Z. and H.W. conceived and designed the experiments; S.B.K. performed the experiments; analyzed the data and wrote the paper; Z.Z. also financed the providing of materials, analysis tools and equipment.

Funding: This research was supported by the Basic Science Center Project of NSFC (Grant No. 51788104), the National Natural Science Foundation of China (Grant No. 51572148 and No. 51531006), and the Tsinghua University Initiative Scientific Research Program.

Conflicts of Interest: The authors declare no conflict of interest. The funding sponsors had no role in the design of the study; in the collection, analyses, or interpretation of data; in the writing of the manuscript, and in the decision to publish the results.

References

1. Sun, J.; Zhang, C.; Zhang, C.; Ding, R.; Xu, Y. Effect of post-treatment on ordered mesoporous silica antireflective coating. *RSC Adv.* **2014**, *4*, 50873–50881. [CrossRef]

2. Zou, L.; Li, X.; Shen, J. Preparation and properties of ordered mesoporous silica antireflective coating with high strength. *Rare Met. Mater. Eng.* **2016**, *45*, 472–476.
3. Xu, L.; He, J. Antifogging and antireflection coatings fabricated by integrating solid and mesoporous silica nanoparticles without any post-treatments. *ACS Appl. Mater. Interfaces* **2012**, *4*, 3293–3299. [CrossRef] [PubMed]
4. Carlstrom, J.E.; Ade, P.A.R.; Aird, K.A.; Benson, B.A.; Bleem, L.E.; Busetti, S.; Chang, C.L.; Chauvin, E.; Cho, H.M.; Crawford, T.M. The 10 meter south pole telescope. *Publ. Astron. Soc. Pac.* **2009**, *123*, 568–581. [CrossRef]
5. Kelzenberg, M.D.; Boettcher, S.W.; Petykiewicz, J.A.; Turnerevans, D.B.; Putnam, M.C.; Warren, E.L.; Spurgeon, J.M.; Briggs, R.M.; Lewis, N.S.; Atwater, H.A. Enhanced absorption and carrier collection in Si wire arrays for photovoltaic applications. *Nat. Mater.* **2010**, *9*, 239–244. [CrossRef] [PubMed]
6. Zou, M.; Thompson, C.; Fleming, R.A. Antireflective Coating for Glass Applications and Method of Forming Same. U.S. Patent WO2014134594A1, 4 September 2014.
7. Moghal, J.; Kobler, J.; Sauer, J.; Best, J.; Gardener, M.; Watt, A.A.R.; Wakefield, G. High-performance, single-layer antireflective optical coatings comprising mesoporous silica nanoparticles. *ACS Appl. Mater. Interfaces* **2011**, *4*, 854–859. [CrossRef] [PubMed]
8. Xu, L.; He, J. A novel precursor-derived one-step growth approach to fabrication of highly antireflective, mechanically robust and self-healing nanoporous silica thin films. *J. Mater. Chem. C* **2013**, *1*, 4655–4662. [CrossRef]
9. Lu, Y.; Zhang, X.; Huang, J.; Li, J.; Wei, T.; Lan, P.; Yang, Y.; Xu, H.; Song, W. Investigation on antireflection coatings for Al:ZnO in silicon thin-film solar cells. *Opt. Int. J. Light Electron. Opt.* **2013**, *124*, 3392–3395. [CrossRef]
10. Wu, F.; Dantan, J.Y.; Etienne, A.; Siadat, A.; Martin, P. Improved algorithm for tolerance allocation based on Monte Carlo simulation and discrete optimization. *Comput. Ind. Eng.* **2009**, *56*, 1402–1413. [CrossRef]
11. Menezes, E.H.S.D.C.; König, P.; Jilavi, M.H.; Oliveira, P.W.D.; Júnior, S.A. Carboxylic acids and esters as scaffold for cavities in porous single layer anti-reflective coatings of silica-titania with excellent optical and mechanical properties. *Mater. Sci. Appl.* **2014**, *5*, 783–788. [CrossRef]
12. Li, T.; He, J. Mechanically robust, humidity-resistant, thermally stable high performance antireflective thin films with reinforcing silicon phosphate centers. *Sol. Energy Mater. Sol. Cells* **2017**, *170*, 95–101. [CrossRef]
13. Khristyan, V.A.; Zagoruiko, Y.A.; Kovalenko, N.O.; Mateychenko, P.V.; Sofronov, D.S. Thermally stable antireflection coatings for active elements of ZnMgSe:Cr^{2+}–laser: Preparation and properties. *Funct. Mater.* **2011**, *18*, 462–465.
14. Doroshenko, M.E.; Osiko, V.V.; Jelínková, H.; Jelínek, M.; Šulc, J.; Němec, M.; Vyhlídal, D.; Čech, M.; Kovalenko, N.O.; Gerasimenko, A.S. Spectroscopic and laser properties of bulk iron doped zinc magnesium selenide Fe:ZnMgSe generating at 4.5–5.1 µm. *Opt. Express* **2016**, *24*, 19824–19834. [CrossRef] [PubMed]
15. Abdul Hadi, S.; Milakovich, T.; Bulsara, M.T.; Saylan, S. Design optimization of single-layer antireflective coating for $GaAs_{1-x}P_x$/Si Tandem Cells With x = 0, 0.17, 0.29, and 0.37. *IEEE J. Photovolt.* **2014**, *5*, 425–431. [CrossRef]
16. Saylan, S.; Milakovich, T.; Hadi, S.A.; Nayfeh, A.; Fitzgerald, E.A.; Dahlem, M.S. Multilayer antireflection coating design for $GaAs_{0.69}P_{0.31}$/Si dual-junction solar cells. *Sol. Energy* **2015**, *122*, 76–86. [CrossRef]
17. Bernal-Correa, R.; Morales-Acevedo, A.; Mora, A.; Pulzara, L.; Monsalve, J.M. Design of $Al_xGa_{1-x}As$/GaAs/$In_yGa_{1-y}As$ triple junction solar cells with anti-reflective coating. *Mater. Sci. Semicond. Process.* **2015**, *37*, 57–61. [CrossRef]
18. Khan, S.B.; Wu, H.; Huai, X.; Zou, S.; Liu, Y.; Zhang, Z. Mechanically robust antireflective coatings. *Nano Res.* **2018**, *11*, 1699–1713. [CrossRef]
19. Kim, S.; Cho, J.; Char, K. Thermally stable antireflective coatings based on nanoporous organosilicate thin films. *Langmuir* **2007**, *23*, 6737–6743. [CrossRef] [PubMed]
20. Wang, S.; Yan, H.; Li, D.; Qiao, L.; Han, S.; Yuan, X.; Liu, W.; Xiang, X.; Zu, X. TEM and STEM studies on the cross-sectional morphologies of Dual-/Tri-layer broadband SiO_2 antireflective films. *Nanoscale Res. Lett.* **2018**, *13*, 49. [CrossRef] [PubMed]
21. Huang, X.; Yuan, Y.; Liu, S.; Wang, W.; Hong, R. One-step sol-gel preparation of hydrophobic antireflective SiO_2 coating on poly(methyl methacrylate) substrate. *Mater. Lett.* **2017**, *208*, 62–64. [CrossRef]

22. Kennedy, S.R.; Brett, M.J. Porous broadband antireflection coating by glancing angle deposition. *Appl. Opt.* **2003**, *42*, 4573–4579. [CrossRef] [PubMed]
23. Li, J.; Lu, Y.; Lan, P.; Zhang, X.; Xu, W.; Tan, R.; Song, W.; Choy, K.L. Design, preparation, and durability of TiO_2/SiO_2 and ZrO_2/SiO_2 double-layer antireflective coatings in crystalline silicon solar modules. *Sol. Energy* **2013**, *89*, 134–142. [CrossRef]
24. Zhao, J.; Green, M.A. Optimized antireflection coatings for high-efficiency silicon solar cells. *IEEE Trans. Electron Devices* **1991**, *38*, 1925–1934. [CrossRef]
25. Zhang, J.; Lan, P.; Li, J.; Xu, H.; Wang, Q.; Zhang, X.; Zheng, L.; Lu, Y.; Dai, N.; Song, W. Sol-gel derived near-UV and visible antireflection coatings from hybridized hollow silica nanospheres. *J. Sol-Gel Sci. Technol.* **2014**, *71*, 267–275. [CrossRef]
26. Du, X.; He, J. Facile fabrication of hollow mesoporous silica nanospheres for superhydrophilic and visible/near-IR antireflection coatings. *Chem. Eur. J.* **2011**, *17*, 8165–8174. [CrossRef] [PubMed]
27. Zhang, L.; Du, K.; Zhou, L.; Tu, H. Preparation of silica antireflective coating for UV-laser. *Acta Opt. Sin.* **1996**, *16*, 998–1001.
28. Belleville, P.F.; Floch, H.G. Ammonia hardening of porous silica antireflective coatings. In *Proc. SPIE 2288, Sol-Gel Optics III, Proceedings of SPIE's 1994 International Symposium on Optics, Imaging, and Instrumentation, San Diego, CA, USA, 24-29 July 1994*; Mackenzie, J.D., Ed.; SPIE: Bellingham, WA, USA, 1994. [CrossRef]
29. Wang, K.; He, J. One-Pot fabrication of antireflective/antibacterial dual-function Ag NP-containing mesoporous silica thin films. *ACS Appl. Mater. Interfaces* **2018**, *13*, 11189–11196. [CrossRef] [PubMed]
30. Mazur, M.; Wojcieszak, D.; Domaradzki, J.; Kaczmarek, D.; Song, S.; Placido, F. TiO_2/SiO_2 multilayer as an antireflective and protective coating deposited by microwave assisted magnetron sputtering. *Opto-Electron. Rev.* **2013**, *21*, 233–238. [CrossRef]
31. Wuu, D.S.; Lin, C.C.; Chen, C.N.; Lee, H.H.; Huang, J.J. Properties of double-layer Al_2O_3/TiO_2 antireflection coatings by liquid phase deposition. *Thin Solid Films* **2015**, *584*, 248–252. [CrossRef]
32. Huang, J.J.; Lee, Y.T. Self-cleaning and antireflection properties of titanium oxide film by liquid phase deposition. *Surf. Coat. Technol.* **2013**, *231*, 257–260. [CrossRef]
33. Kuo, C.F.J.; Tu, H.M.; Su, T.L. Optimization of the electron-beam-lithography parameters for the moth-eye effects of an antireflection matrix structure. *J. Appl. Polym. Sci.* **2006**, *102*, 5303–5313. [CrossRef]
34. Khan, S.B.; Wu, H.; Li, J.; Chen, L.; Zhang, Z. Bilayer SiO_2 nanorod arrays as omnidirectional and thermally stable antireflective coating. *Adv. Eng. Mater.* **2018**, *20*, 1700942. [CrossRef]
35. Vicente, G.S.; Bayón, R.; Germán, N.; Morales, A. Long-term durability of sol–gel porous coatings for solar glass covers. *Thin Solid Films* **2009**, *517*, 3157–3160. [CrossRef]
36. Khan, S.B.; Wu, H.; Ma, L.; Hou, M.; Zhang, Z. HfO_2 nanorod array as high-performance and high-temperature antireflective coating. *Adv. Mater. Interfaces* **2017**, *4*, 1600892. [CrossRef]
37. Shirolkar, M.; Kazemian Abyaneh, M.; Singh, A.; Tomer, A.; Choudhary, R.; Sathe, V.; Phase, D.; Kulkarni, S. Rapidly switched wettability of titania films deposited by dc magnetron sputtering. *J. Phys. D Appl. Phys.* **2008**, *41*, 1525–1528. [CrossRef]
38. Takeda, S.; Suzuki, S.; Odaka, H.; Hosono, H. Photocatalytic TiO_2 thin film deposited onto glass by DC magnetron sputtering. *Thin Solid Films* **2001**, *392*, 338–344. [CrossRef]
39. Pérez-González, M.; Tomás, S.A.; Santoyo-Salazar, J.; Morales-Luna, M. Enhanced photocatalytic activity of TiO_2–ZnO thin films deposited by dc reactive magnetron sputtering. *Ceram. Int.* **2017**, *12*, 8831–8838. [CrossRef]
40. Pérez-González, M.; Tomás, S.A.; Morales-Luna, M.; Arvizu, M.A.; Tellez-Cruz, M.M. Optical, structural, and morphological properties of photocatalytic TiO_2–ZnO thin films synthesized by the sol–gel process. *Thin Solid Films* **2015**, *594*, 304–309. [CrossRef]

© 2018 by the authors. Licensee MDPI, Basel, Switzerland. This article is an open access article distributed under the terms and conditions of the Creative Commons Attribution (CC BY) license (http://creativecommons.org/licenses/by/4.0/).

Article

Stabilized SPEEK Membranes with a High Degree of Sulfonation for Enthalpy Heat Exchangers

Riccardo Narducci [1,2,*], Maria Luisa Di Vona [1,2], Assunta Marrocchi [3] and Giorgio Baldinelli [4]

1 Department of Industrial Engineering, University of Rome Tor Vergata, 00100 Roma, Italy; divona@uniroma2.it
2 International Associated Laboratory (L.I.A.), Ionomer Materials for Energy, 00133 Roma, Italy
3 Department of Chemistry, Biology and Biotechnology, University of Perugia, 06123 Perugia, Italy; assunta.marrocchi@unipg.it
4 Department of Engineering, University of Perugia, 06125 Perugia, Italy; giorgio.baldinelli@unipg.it
* Correspondence: riccardo.narducci@uniroma2.it; Tel.: +39-06-7259-4388

Received: 18 April 2018; Accepted: 16 May 2018; Published: 19 May 2018

Abstract: In this investigation, we explored for the first time the use of cross-linked sulfonated poly (ether ether ketone) (SPEEK) membranes in the fabrication of enthalpy heat exchangers. SPEEK is very sensitive to changes in relative humidity, especially when featuring high degrees of sulfonation (DS), though a poor mechanical stability may be observed in the latter case. Cross-linking is crucial in overcoming this issue, and here, we firstly employed the INCA method (ionomer counter-elastic pressure "n_c" analysis) to assess the improvements in the mechanical properties. The cross-link was achieved following a simple thermal-assisted process that occurs directly on the performed membranes. After an initial screening, a degree of cross-link = 0.1 was selected as the better compromise between absorption of water vapor and mechanical properties. When implemented in the enthalpy heat exchanger system, these cross-linked SPEEK membranes enabled a high level of sensible heat exchange, as well as a remarkable variation in the mass (water vapor) transfer between the individual air flows. The performances resulted in being better than those for the system based on a benchmark commercially available perfluorinated Nafion membrane.

Keywords: HVAC; SPEEK; cross-linking; INCA method; thin membranes; high DS

1. Introduction

The impelling needs for a consistent reduction of pollution in large towns and CO_2 emissions in the atmosphere have reinforced the interest in efficient and clean systems for homes and other buildings. In terms of the evolution of established policies, in 2014, the European Union issued Regulation 1253/2014 on eco-design requirements for ventilation units, which represents one of the measures implementing Directive 2009/125/EC, establishing a framework for the settings of design specifications for improving the environmental performance of energy-related products. In this broad context, the building sector is therefore pushed ever more towards reducing energy consumption, leading to the consequent need for improving the "indoor" insulation. On the other hand, insulation may lead to an increase in air tightness, thereby causing poor indoor environments and causing adverse effects on buildings and users. To avoid negative effects due to the lack of the air exchange, several companies proposed controlled mechanical ventilation systems, the result of which is usually associated with a heat recovery system (so-called exchanger or heat recuperator) to increase the energy efficiency of the ventilation system itself. The thermal load for ventilation systems now represents a relevant percentage of the overall energy needs, and therefore, any direct effort towards the increase in efficiency of the ventilation itself would bring significant benefits for the whole building sector. Through the installation of the aforementioned systems, it is possible to avoid the arbitrary opening action of the fixtures, pursuing the goal of energy saving by preheating the outside cold air with warm internal air at the outlet (in winter conditions) or pre-cooling the warm outdoor air with cold indoor air at the exhaust (in summer conditions). In the last few years, many systems have emerged that allow, in addition to the heat recovery, to humidify (in winter) and dehumidify (in summer) the outside air [1]. These systems are called enthalpy heat exchangers. The recovery of water vapor is particularly useful from an energy point of view, as it contributes to passively decreasing the energetic load that the air conditioning plants must support to remove water vapor in the warm season and to increase its presence in the cold season. Furthermore, the enthalpy heat exchangers show themselves to be useful for improving the indoor thermohygrometric comfort [2]. Though their structure is often rather simple, the practical applications of the enthalpy exchangers are limited. This is essentially related to the very low moisture transfer coefficient of the common vapor-permeable materials with respect to the associated cost increase, which ultimately lead to an energy recovery device with a long payback time. Most enthalpy heat exchangers make use of a membrane involving the exchange of heat while being permeable to water to guarantee the water vapor transfer [3–5]. Among these, ion-exchange membranes (IEM) are particularly attractive because of their effective water transport ability, which is well demonstrated in different fields, including fuel cells, flow batteries, water purification, pharmaceutical industry and the food industry [6–8]. Despite this fact, there is only a handful of studies in this area, mostly patented [9–12]. Perfluorinated (IEM) materials like Nafion™ [13–18] appear to be the most obvious choice, though they present important drawbacks. In 2006, the U.S. Environmental Protection Agency (EPA) demonstrated the bioaccumulative effects of several perfluoro-organic compounds featuring long perfluoro alkyl chains [19]. Many of these compounds are currently classified as substances of very high concern (SHVC) under EU Chemicals Regulation REACH [20]. Therefore, companies involved in the manufacture and marketing of fluoro-compounds are replacing long perfluorinated alkyl compounds with shorter ones. However, the shortening of the chain length of the fluorinated moiety has posed challenging technological issues because of the dramatic loss of performances such as a high degree of swelling in the presence of moisture, with consequent deterioration of mechanical properties, up to breakage [21]. The loss of performance may ultimately lead to a decrease in the heat and water vapor exchange capacity. Another limit of perfluorinated materials is their high cost (i.e., U.S. dollars 140 to 311/0.305 m × 0.305 m for Nafion™, Ion Power) [22].

Among all the candidates for IEMs, sulfonated poly (ether ether ketone)-based (SPEEK, Figure 1) membranes are considered to be the most attractive non-perfluorinated ones [22–25].

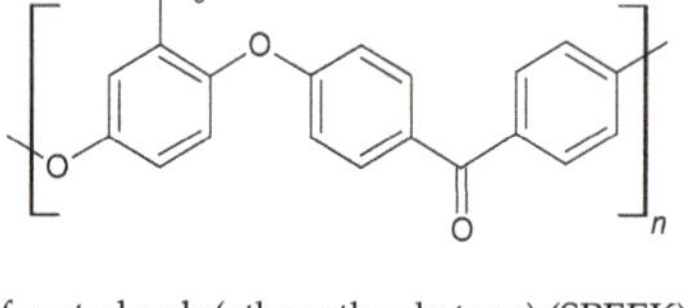

Figure 1. Sulfonated poly(ether ether ketone) (SPEEK) repeat unit.

They combine appealing properties, including high thermal and chemical stability, easy availability and low cost; on the other hand, SPEEK membranes with a high sulfonation degree (DS) can dissolve in liquid water or in particular conditions of temperature and relative humidity (RH), completely losing their mechanical properties. In 2009, some of us contributed to the development of simple and cost-effective thermal cross-linking treatments assisted by the presence of a polar aprotic solvent, i.e., dimethyl sulfoxide (DMSO), to enhance the stability and performance of SPEEK membranes. It was demonstrated that after treatments performed in air at temperatures higher than 160 °C, SPEEK membranes have a Young modulus higher (1300 MPa) than the untreated counterpart (850 MPa) and could resist water up to 145 °C without significant swelling [26–28]. Surprisingly, the cross-linking does not decrease the capacity of the polymer to absorb water vapor, due to a reduced tortuosity of the membrane channels, enabling a better water transport in the membrane microstructure [29]. Note that with respect to the known cross-linking processes [30–38], the above approach does not use cross-linking molecules, which are often sensitive to the operating conditions [39]. Moreover, the process as a whole is not-expensive, short-lived, simply realizable and is suitable for all DS values [40–46].

In this paper, we explore for the first time the effectiveness of cross-linked sulfonated SPEEK membranes in the fabrication of enthalpy heat exchangers. Our results demonstrated that a high level of sensible heat exchanged (the energy transfer linked to temperature variations), as well as a remarkable variation in the mass (water vapor) transfer between the individual air flows are achievable. Commercially available Nafion™ NRE-212 was also considered for comparative purposes.

2. Materials and Methods

2.1. Materials

PEEK (Victrex 450P, molecular weight MW = 38,300 g/mol, 132 repeat units per mole) was supplied by Solvay (Brussels, Belgium). Nafion™ NRE-212 (equivalent weight EW = 1100 g/eq, ion exchange capacity (IEC) = 0.91 meq/g) were supplied by Sigma-Aldrich (St. Louis, MO, USA). Dimethyl sulfoxide (DMSO) and all the other reagents and materials were purchased from Carlo Erba RP (Milan, Italy).

2.2. Membrane Preparation

SPEEK was prepared by the reaction of poly(ether ether ketone) (PEEK) with concentrated sulfuric acid (95%–98%), under N_2 at 50 °C for 2–4 days, following a previously-reported procedure [47]. A DS in the range 0.7–1.0, depending on the reaction time, was obtained. The solution was poured, under continuous stirring, into an excess of ice-cold water obtaining a white precipitate. After resting overnight, the precipitate was filtered and washed with water several times, using a dialysis membrane (Sigma–Aldrich D9402), to neutral pH to eliminate the residual sulfuric acid completely. The sulfonated polymer (SPEEK) was then dried over night at 80 °C. The polymers were dissolved in dimethyl sulfoxide (DMSO) at 80–90 °C. The ratio polymer: DMSO was 1:10 (mg/mL). After evaporation to ~1/3 of the original volume, the solution was spread on a glass plate, using doctor blade-type equipment and then heated in the oven for the casting treatment at different times (15–24 h) and temperatures (90–120 °C). After the casting, membranes were stored at ambient relative humidity and peeled off. Membranes treated at 120 °C for 24 h (in the following 120–24) were placed on a Teflon substrate and put in the ventilated oven at 180 °C (cross-linking procedure, called in the following XL SPEEK) for different times in the range 3–24 h depending on the desired crosslinking degree (DXL) [27,40]. After the preparation, all the membranes were immersed at room temperature in a solution of H_2O_2 3% for 1 h, then in H_2SO_4 5 M for 1 h and rinsed off with water. The membrane thickness was 20 µm.

2.3. Membrane Characterization

2.3.1. Ion Exchange Capacity Measurements

The ion exchange capacity (IEC in milliequivalents/gram) was measured by titration [27]. To eliminate DMSO, which can affect IEC, membranes were swelled in water at 100 °C for 5 h, followed by treatment in H_2SO_4 5 M at room temperature for 2 h and then washed until neutral pH before the titration.

The degree of crosslinking (DXL) for treated samples was evaluated measuring the IEC using the following formula:

$$DXL = (IEC_{init} - IEC_{fin})/IEC_{init} \qquad (1)$$

where IEC_{init} refers to the ion exchange capacity before the treatments and IEC_{fin} refers to the ion exchange capacity after the treatments at 180 °C.

2.3.2. Water Uptake

Membrane samples dried over P_2O_5 for 3 days were weighed m_{dry} and then immersed for 24 h in liquid water in a closed Teflon vessel at a constant temperature. After the immersion, the membranes were equilibrated at 25 °C in water for 24 h. The excess of water was carefully wiped off, and the membrane mass was determined m_{wet}:

$$WU = \frac{m_{wet} - m_{dry}}{m_{dry}} \times 100 \qquad (2)$$

The hydration number λ was calculated as:

$$\lambda = \frac{n(H_2O)}{n(SO_3H)} = \frac{WU}{IEC \times M(H_2O)} \times 1000 \qquad (3)$$

where n is the mol of water per mol of the sulfonic group, M the molar mass of water and IEC the ion exchange capacity. The uncertainty in the measurements was ~±0.5.

2.3.3. Counter-Elastic Pressure (n_c) Measurements

The n_c index was obtained from water-uptake after equilibration in distilled liquid water at room temperature for 24 h [21,48]. The λ values were converted into n_c values by Equation (4) [13,49]:

$$n_c = \frac{100}{\lambda - 6} \qquad (4)$$

2.3.4. Membrane Density, Volume

A portion of the membrane (~15 mm × 15 mm) was first dried for one day over P_2O_5 at room temperature and then the weight and the a, b and c dimensions accurately determined by a micrometer. The dried membrane was immersed in bi-distilled water at 20 °C and equilibrated for 24 h, then the weight and a, b and c dimensions were again determined [14,48].

2.3.5. DMA

The dynamic mechanical analysis (DMA) analysis was performed from 30–250 °C in air on the DMA 2980 dynamic analyzer (TA Instrument, New Castle, DE, USA). The frequency was fixed (1 Hz) with 0.05 N initial static force and an oscillating amplitude of 10 μm [40].

2.3.6. Enthalpy Heat Exchanger

A simplified model of the basic unit of the enthalpy heat exchanger has been constructed (Figure 2, experimental apparatus, and Figure 3, scheme of the enthalpy heat exchanger of the experimental apparatus). The main components of the test bench are shown in Figure 4: the humidification chamber (a), which consists of a Pexiglass vessel measuring 50 cm × 50 cm × 50 cm, with an opening side for the positioning of the suction fans (b) and connected, through a soft mobile tube, to an ultrasonic humidifier, which allows one to control the temperature and relative humidity conditions at the inlet of the exchanger (c), varying them within the characteristic intervals of the applications related to the air treatment in buildings [50]. The membranes used are: plastic materials, Nafion NRE™ 212, XL SPEEK (DXL = 0.1, DS_{init} = 0.9).

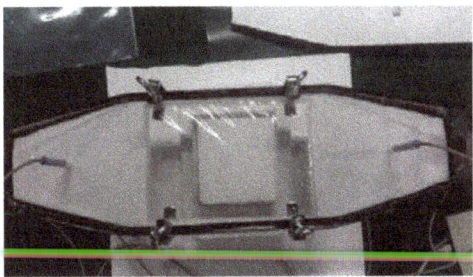

Figure 2. Support including the membrane for the base unit (plate) of the enthalpy heat exchanger.

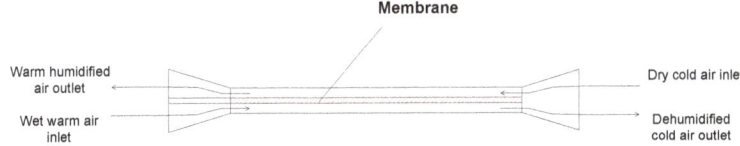

Figure 3. Schematic representation of the enthalpy heat exchanger used for the experimental test.

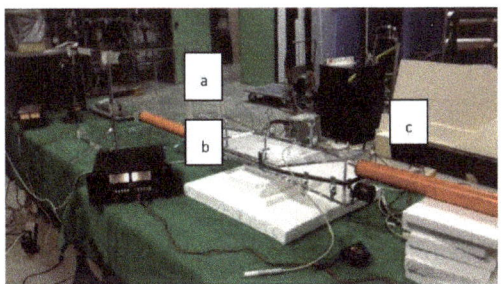

Figure 4. Test bench for the performance measurement of the basic unit of the enthalpy heat exchanger. The humidification chamber (**a**), suction fans (**b**), exchanger (**c**).

3. Results and Discussion

The measurements on the test bench described in the Materials and Methods were performed on XL SPEEK membranes due to the instability of membranes without treatment, as explained below. The water uptake (WU) was measured at different temperatures and times in liquid water on SPEEK membranes with DS = 0.70 and 0.9 treated in oven at 90 °C for 24 h (90–24) or 120 °C for 15 h (120–15). The kinetics of the water uptake and the time where the equilibrium is reached depend very much on the DS. For SPEEK with DS = 0.70, WU kinetics were made at different temperatures from 1.5–15 °C; at 20 °C, the material begins to show an excessive swelling. It is not possible to perform kinetic measurements in the same conditions for SPEEK with DS = 0.9 (90–24), because of the solubilization of the material after 5 h. Figure 5 shows that for SPEEK 0.70 (90–24), the water uptake kinetics at 1.5 °C is very fast and reached the stationary value after only 5 h of hydration with a final WU = 70% corresponding to a λ value of 20 (Table 1). The material reaches equilibrium rapidly (5–10 h) at low temperatures, while the equilibrium is shifted to larger times with increasing temperature (at 15 °C around 40 h). This behavior is similar to that observed for Nafion NRE-212 (EW 1100), where the material is subjected to two processes: the first one is rapid, and it is due to the diffusion of water inside the membranes; the second one is slower, and it is related to irreversible changes of chain conformations [50].

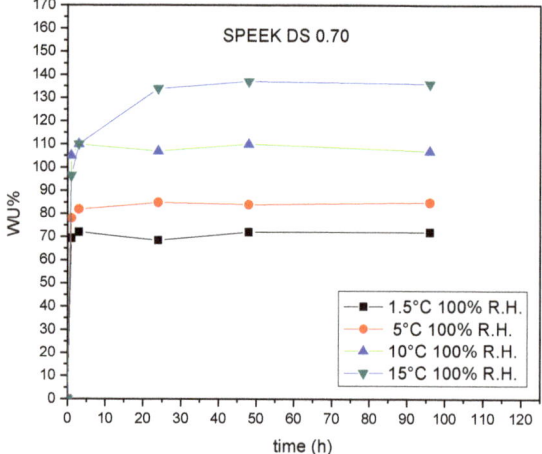

Figure 5. Water uptake (WU) vs. time of SPEEK DS 0.70 (90–24).

Table 1. Water uptake (WU), lambda (λ) and counter-elastic pressure (n_c) for SPEEK DS 0.7 (90–24) in liquid water at equilibrium.

Temperature (°C)	WU (%) H$_2$O 24 h (±5%)	λ H$_2$O 24 h (±5%)	n_c H$_2$O 24 h (±5%)
1.5	70	20	7.1
5	85	23	5.9
10	107	29	4.2
15	135	32	3.8

A heat treatment at 120 °C for 15 h was performed in order to decrease the free volume between the chains, for the improvement of the hydrophilic stability of SPEEK with DS = 0.9, which presents a large swelling even at a low temperature. After the treatment, the material begins to solubilize at 10 °C and no more at 1 °C, due to the annealing effect (Figure 6 and Table 2).

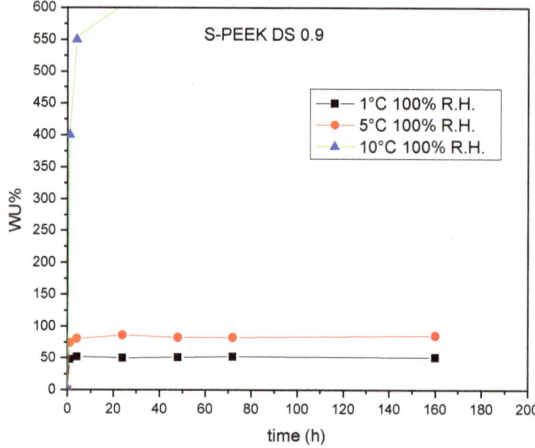

Figure 6. Water uptake vs. time for SPEEK DS 0.9 (120–15) in liquid water.

Table 2. Water uptake (WU), lambda (λ) and counter-elastic pressure (n_c) for SPEEK DS 0.9 (120–15) in liquid water at equilibrium.

Temperature (°C)	WU (%) H$_2$O 24 h (±5%)	λ H$_2$O 24 h (±5%)	n_c H$_2$O 24 h (±5%)
1	50	12	15.4
5	80	20	7.1
10	600	150	0.7

Figure 7 shows lambda (λ) values for SPEEK with DXL = 0.22 measured after swelling in water at 100 °C for different times and equilibrated at room temperature for 24 h. We can recognize two processes: an initial and fast one and a slower one. The latter is probably due to the slow morphological change of the backbone.

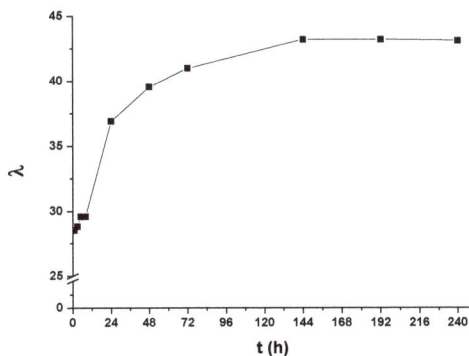

Figure 7. Equilibrium water uptake coefficients (λ) vs. time for a sample treated at 120–24 and 180–10 with final DS = 0.73, ion exchange capacity (IEC) = 2.1 and crosslinking degree (DXL) = 0.22.

In Figure 8 is reported an application of the INCA method (ionomer n_c analysis). The INCA method is based on the use of n_c/T plots to derive important properties of the ionomer, such as the degree of crystallinity, mechanical properties, glass transition and melting temperature. The n_c vs. temperature graph shows that membranes completely lose their mechanical properties (n_c = 0), and they present infinite swelling already at 10 °C for SPEEK with DS 0.9 and at 25 °C with DS 0.7. The weak van deer Waals forces inside the macromolecular chains, which are broken at low temperatures, and the presence of a larger number of sulfonic groups, which determine a greater absorption of water, are responsible for the previous behavior [21,49]. The presence of covalent bonds in the XL derivative (Figure 9) completely changes this trend, improving the resistance in water.

As already discussed before, the n_c analysis allowed evaluating the mechanical properties. A comparison of the results was also made using stress strain tests. The T_g values were determined with the dynamic mechanical analysis (DMA). It can be seen that the intrinsic values of T_g of SPEEK are very high and therefore suitable for use in all environmental conditions. The results are shown in Table 3.

In general, Young's modulus (E, stiffness) characterizes the elastic domain of polymers, where weak inter-chain bonds are observed at the microscopic scale; this could be related to fundamental bond properties. Stiffness explores essentially low displacements such as (a) van der Waals bonds (the change of the distance between chains and of dipole-dipole interactions), (b) defects, such as entanglements, and (c) the presence of solvents, such as water and DMSO. Yield stress (YS) and tensile strength (TS) are instead related to strong bonds, including covalent cross-links between macromolecular chains, and they constitute macroscopic scale properties. Tensile strength and yield stress explore large displacements (plasticity). It is evident from the mechanical measurements that the thermal treatment at 180 °C enhances significantly the membrane mechanical properties: the elastic modulus, the ultimate tensile strength and the yield stress all increase significantly by XL increasing. It can be seen that the elongation at break of the membrane with DXL = 0.10 is good and keeps a value close to the SPEEK with DXL = 0 (ductile behavior), which corresponds to a greater plasticity and malleability of the material. For greater degrees of XL, the value is typical of brittle polymers. If we assume that the mechanical degradation of membranes is related to the existence of plastic deformation during the operation, the enhancement of mechanical properties is of major importance for the improvement of membrane durability. Table 4 reports the WU and the swelling parameters for SPEEK 0.9. Only the XL treatment allows avoiding the complete dissolution of the material.

Figure 8. Plot n_c/T of SPEEK in liquid water: SPEEK DS 0.9, blue triangles; SPEEK DS 0.7, green triangles; XL SPEEK DS 0.9 (DXL = 0.1), red triangles.

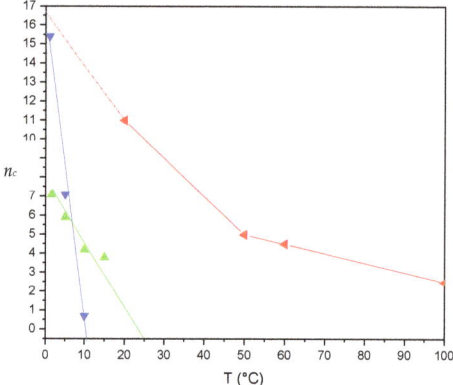

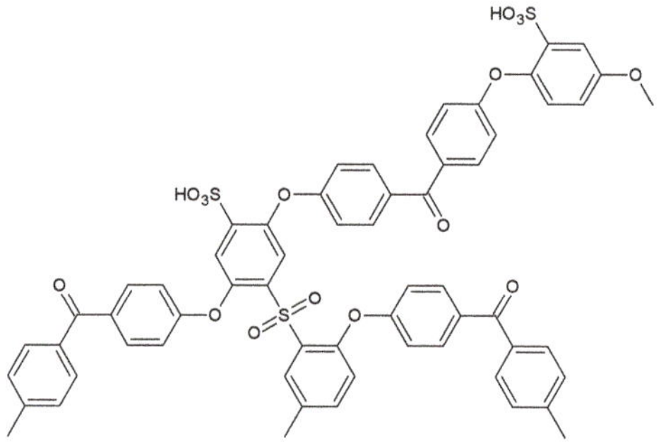

Figure 9. Repeat unit of SPEEK containing sulfonic groups and sulfone bridges.

Table 3. Mechanical properties (glass transition temperatures (T_g) [40] for various SPEEK membranes in acid form with an initial IEC= 2.5 meq/g, DS = 0.9 [51] and E: elastic modulus, UTS: ultimate tensile strength, YS: yield stress, ε: elongation at break [52]).

SPEEK	DXL	E/MPa	UTS/MPa	YS/MPa	ε	T_g/°C
120 °C, 24 h	0	850 ± 60	32 ± 1	20 ± 2	20 ± 8	180
180 °C, 3 h	0.1	1160 ± 50	41 ± 2	30 ± 2	22 ± 9	193
180 °C, 10 h	0.22	1300 ± 100	43 ± 8	35 ± 5	20 ± 12	239
180 °C, 24 h	0.35	1450 ± 50	59 ± 2	45 ± 2	11 ± 1	>250

Table 4. Water uptake (WU), lambda (λ), counter-elastic pressure (n_c), volume wet (V_{wet}), and density wet (d_{wet}) were measured in liquid water for 24 h at 25 °C; volume dry (V_{dry}) and density dry (d_{dry}) were measured under P_2O_5 for 24 h for SPEEK (DS = 0.9) and XL SPEEK (DXL = 0.16) [50].

Sample	WU (%) (±5%)	λ (±5%)	n_c (±5%)	V_{wet} (mm³)	d_{wet} (g/cm³)	V_{dry} (mm³)	d_{dry} (g/cm³) (±5%)
SPEEK 120–24 (DXL = 0)	∞	∞	0	–	–	25.4	1.35
XL-SPEEK 180-7 H_2O 100 °C (DXL = 0.16)	100	25	5	56.8	1.22	24.8	1.38

Due to the presence of the XL, it is possible to use high values of DS that allow the material to acquire a great sensitivity to changes in relative humidity, since the tendency to absorb water is proportional to the number of sulfonic groups per repeating unit. From the point of view of applications in heat exchangers, it is interesting to note that the capacity to adsorb water does not change significantly until RH = 80% for XL and non-XL derivatives (Table 5).

Table 5. Environmental humidity (RH) and water uptake (WU) of SPEEK polymer with different degrees of cross-linking (DXL) [53].

DXL = 0		DXL = 0.1		DXL = 0.22	
RH (±1%)	WU (±5%)	RH (±1%)	WU (±5%)	RH (±1%)	WU (±5%)
15	2	15	2	15	2
40	8	40	7	40	8
58	10	58	10	58	12
80	21	80	18	80	23
97	46	97	36	97	30

The effectiveness of SPEEK membranes for heat exchangers was explored on a simplified geometry (flat plate) (see Materials and Methods). In the proposed experimental apparatus (Figure 4), it is therefore possible to fix the temperature and relative humidity conditions of the two air flows, controlling also the flow rate, with the aim of evaluating the capacity of membranes to transfer heat and humidity. The operating scheme of the test bench is as follows: in the humidification chamber, humid and warm air are produced to simulate the outdoor air conditions in the summer. The flow of the air drawn from the said chamber will constitute the air flow rate that will have to exchange heat and water vapor with the dry and wet flow coming from the inlet. The two flows are sent to the exchanger and, excluding the initial part of the convergent, will carry out the heat and mass exchange, mainly in correspondence with the membrane under examination. The polymer membrane is therefore lapped on both sides by the two flows at known temperature and relative humidity, measured at the inlet and at the outlet of the heat exchanger itself through four sensors (Galltec-Mela, L series). Data are acquired through a program developed in the Labview environment, which allows saving, processing and graphic representation. From the physical point of view, the heat given off by the hot fluid will be given by its thermal capacity for the decrease in temperature that it undergoes along the passage on the membrane; the water vapor yielded will instead be given by the reduction in absolute humidity contained in the air flow (absolute humidity is the ratio between the mass of water and the mass of dry air contained in a given volume of humid air). The opposite will happen for cold and dry fluid. The relative humidity is, together with the temperature, the thermohygrometric parameter most closely linked to the well-being of people in the indoor environment and is also the simplest parameter to measure with the available instruments. Since the temperature and relative humidity of air can be traced back to its water content, analyses of temperature variation and relative humidity in the test bench were carried out to measure the performance of the exchanger. The verification of the thermohygrometric exchange efficiency can be carried out both through the separate evaluation of temperature and relative humidity variations. The results are shown in Figures 10–13. The test bench was tested with a sheet of waterproof plastic material, in order to validate the experimental apparatus. The first experimental test was carried out with a flow of hot (31.4 °C) and humid air (relative humidity 95%) at the entrance (green line of Figure 10, red line of Figure 11 and point P1 of Figure 12) and a corresponding flow of cold (26.3 °C) and dry (38% relative humidity) air entering from the opposite side of the heat exchanger (gray line of Figure 10, light blue line of Figure 11 and point P1 of Figure 13) [50].

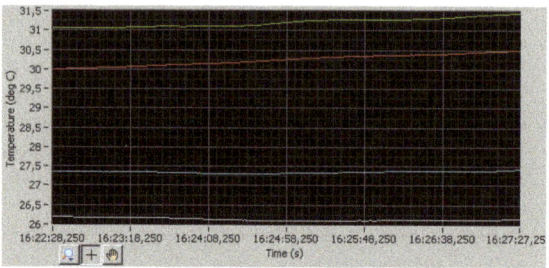

Figure 10. The graph representing the variation of the temperature of the hot stream (green input and red output) and cold flow (gray input and light blue output).

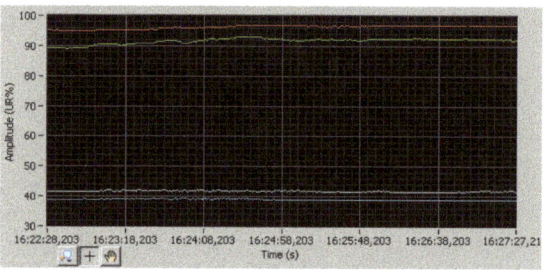

Figure 11. The graph representing the variation of the relative humidity of the wet stream (entry in red and exit in green) and of the dry stream (entry in light blue and exit in gray).

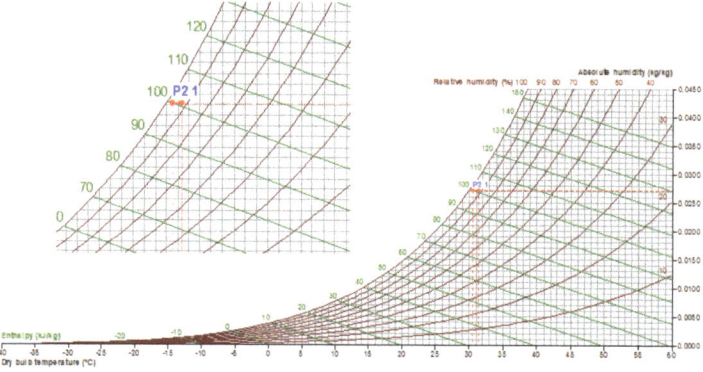

Figure 12. Psychrometric diagram representing the variation of the thermodynamic state of the hot and humid flow in a plastic material.

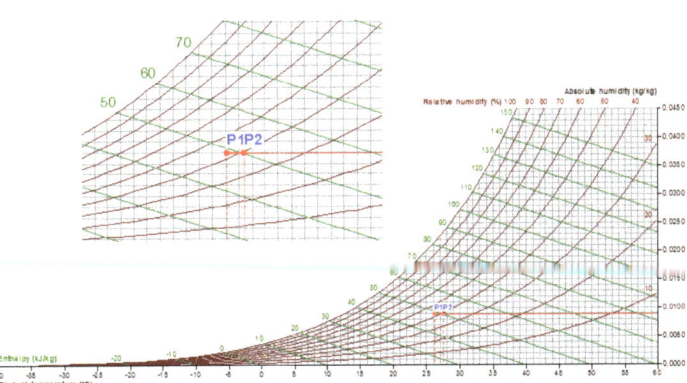

Figure 13. Psychometric diagram representing the variation of the thermohygrometric state of cold and dry flow in a plastic material.

Changes in temperature and water content can be viewed even more immediately through the use of the psychrometric diagram (Figures 12 and 13). In this kind of graph, the temperature of the humid air is shown in the abscissa, in the right-hand ordinate the absolute humidity (in g of water/kg of dry air) and in the curves (in red) running from the bottom side to left to the upper right part, the relative humidity. By fixing two of these three parameters, the thermodynamic status of the humid air can be defined. Thanks to the results, it can be seen how the plastic material allows, as expected, the transfer of sensible heat between the two flows (section P1–P2 of the curves in Figures 12 and 13, which testify to the temperature variations), without allowing the exchange of the specific humidity (both P1–P2 sections are horizontal, that is at constant absolute humidity) [50].

The test was repeated under similar conditions with a membrane of Nafion™ NRE-212 as a comparative reference (Figures 14 and 15). In addition to heat transfer (temperature variations), it is also possible to highlight the transport of water vapor, by modifying absolute humidity in the P1–P2 section of the curves (increasing in the dry stream, decreasing in the wet stream). As an example, the wet flow passes, from 0.0260–0.0245 g/kg, with an absolute humidity decrease equal to 0.0015 g/kg [50].

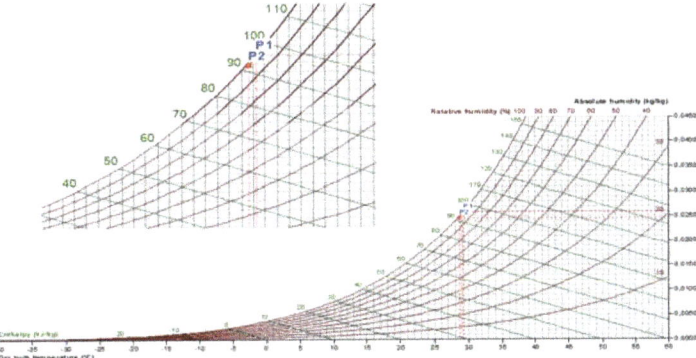

Figure 14. Psychometric diagram representing the variation of the thermodynamic state of the hot and humid flow in Nafion™ NRE-212.

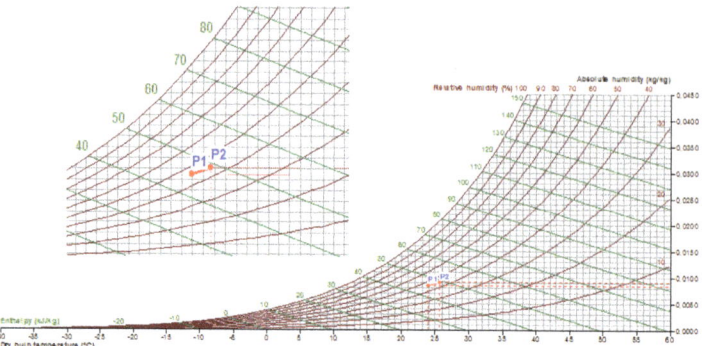

Figure 15. Psychrometric diagram representing the variation of the thermohygrometric state of the cold and dry flow in Nafion™ NRE-212.

The test was finally repeated with a low cross-linked SPEEK-based membrane (DXL = 0.1, DS_{init} = 0.9, IEC_{init} = 2.5 meq/g, 20 µm). This XL value represents the best compromise between water uptake and mechanical properties. The short treatment time (3 h) gives the possibility to produce XL-SPEEK at an industrial scale, where in addition to reliability and performance, the processing time is very important. Figures 16 and 17 show the results obtained. The graphs show that in the passage through the SPEEK membrane, the two air flows exchanged sensible heat, as the temperature variations indicate. Changes in absolute humidity testify, on the other hand, the exchange of mass (water vapor) between the two flows through the membrane, according to the equations previously described. The effectiveness of SPEEK as a membrane in enthalpy heat exchangers is therefore demonstrated. Moreover, it is possible to observe the higher slope of the P1–P2 sections in the tests carried out with cross-linked SPEEK, with respect to the slope of the P1–P2 sections of Nafion ™ with EW = 1100 g/mol, resulting in a more marked variation of the content of water vapor for the individual air flows, so demonstrating the higher performance of SPEEK in terms of water vapor exchange. In fact, the wet flow in this case passes, from 0.0250–0.0230 g/kg, with an absolute humidity decrease equal to 0.0020 g/kg (0.0015 g/kg for Nafion™) [50].

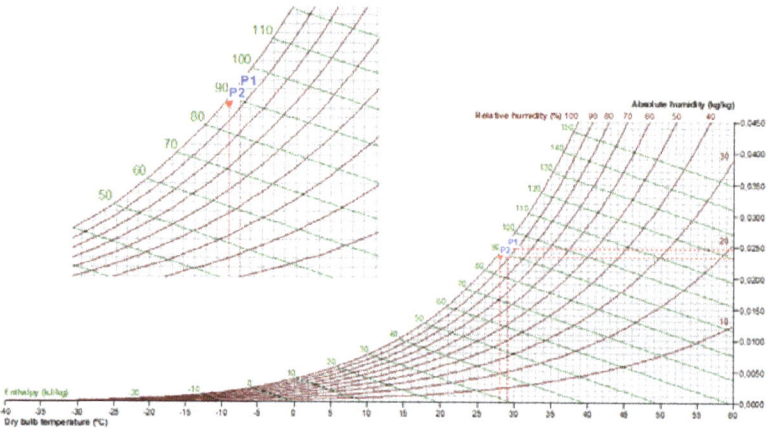

Figure 16. Psychrometric diagram representing the variation of the thermodynamic state of hot and humid flow in crosslinked SPEEK (DXL = 0.1).

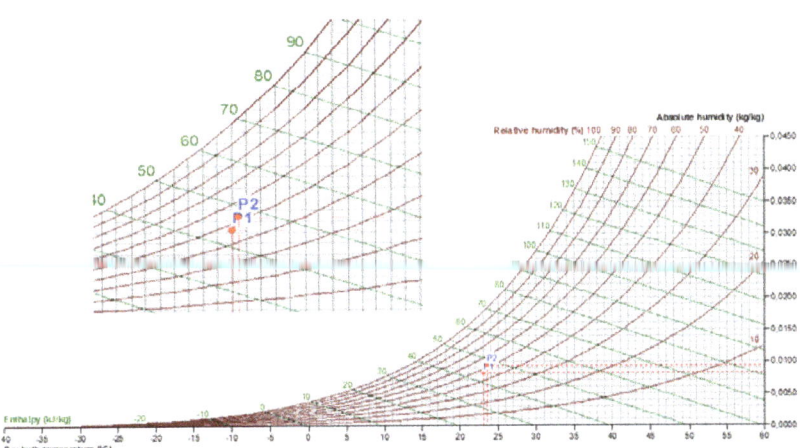

Figure 17. Psychrometric diagram representing the variation of the thermo-hygrometric state of cold and dry flow in cross-linked SPEEK (DXL = 0.1).

4. Conclusions

In this paper, the use of cross-linked SPEEK membranes in the realization of enthalpy heat exchanger systems was proposed for the first time. SPEEK with DS = (0.7–0.9) has very high water uptake with very fast kinetics. Unfortunately, for use in HVAC systems, it is not suitable due to the insufficient mechanical properties ($n_c = 0$, INCA method) even at low temperatures and high humidity values, conditions that usually occur. Thanks to our cross-linking method, we were able to stabilize these materials modulating the grade of XL with different times of treatment. SPEEK membranes were tested in acid form with an initial IEC_{init} = 2.5 meq/g (DS_{init} = 0.9) and a time of treatment of 3 h at 180 °C, corresponding to a DXL = 0.1. These conditions of use represent the best compromise between the water uptake and the mechanical properties. By this design concept, the enthalpy exchangers based on cross-linked SPEEK showed a higher slope of the P1–P2 sections, with respect to the slope of the P1–P2 sections of Nafion™, resulting in a more marked variation of the content of water vapor for the individual air flows. The wet flow in this case passes, from 0.0250–0.0230 g/kg, with an absolute humidity decrease equal to 0.0020 g/kg (0.0015 g/kg for Nafion™). These results are very promising for the utilization of XL SPEEK for enthalpy heat exchangers, in air conditioning applications. The future work will deal with the optimization of membrane shaping to form channels for the enthalpy heat exchangers, both in terms of the geometric section and patterns of air flow.

5. Patents

Patent pending: Di Vona, M.L.; Baldinelli, G.; Marrocchi, A.; Narducci, R. Scambiatori di Calore Entalpici a membrane di Tipo Polimerico Aromatic Solfonato e Procedimento per la Preparazione di Dette Membrane. Domanda numero: 102016000112268, 8 November 2016. (In Italian).

Author Contributions: R.N. and G.B. conceived of, designed and performed the experiments. R.N., G.B., A.M. and M.L.D.V. analyzed the data. G.B., A.M. and M.L.D.V. contributed reagents/materials/analysis tools. R.N. wrote the paper.

Funding: This research received no external funding.

Conflicts of Interest: The authors declare no conflict of interest.

References

1. Reay, D.A. *Heat Recovery Systems*; E & F.N. Span: London, UK, 1979.
2. Fernandez-Seara, J.; Diz, R.; Uhia, F.J.; Dopazo, A.; Ferro, J.M. Experimental analysis of an air-to-air heat recovery unit for balanced ventilation systems in residential buildings. *Energy Convers. Manag.* **2011**, *52*, 635–640. [CrossRef]
3. Al-Waked, R.; Nasif, M.S.; Morrison, G.; Behnia, M. CFD simulation of air to air enthalpy heat exchanger: Variable membrane moisture resistance. *Appl. Therm. Eng.* **2015**, *84*, 301–309. [CrossRef]
4. Woods, J. Membrane processes for heating, ventilation, and air conditioning. *Renew. Sustain. Energy Rev.* **2014**, *33*, 290–304. [CrossRef]
5. Zhang, L.Z.; Liang, C.H.; Pei, L.X. Heat and moisture transfer in application scale parallel-plates enthalpy exchangers with novel membrane materials. *J. Membr. Sci.* **2008**, *325*, 672–682. [CrossRef]
6. Xu, T.W. Ion exchange membranes: State of their development and perspective. *J. Membr. Sci.* **2005**, *263*, 1–29. [CrossRef]
7. Daufin, G.; Escudier, J.P.; Carrere, H.; Berot, S.; Fillaudeau, L.; Decloux, M. Recent and emerging applications of membrane processes in the food and dairy industry. *Food Bioprod. Process.* **2001**, *79*, 89–102. [CrossRef]
8. Tarvainen, T.; Svarfvar, B.; Akerman, S.; Savolainen, J.; Karhu, M.; Paronen, P.; Jarvinen, K. Drug release from a porous ion-exchange membrane in vitro. *Biomaterials* **1999**, *20*, 2177–2183. [CrossRef]
9. Dobbs, G.M.; Freihaut, J.D. Plate-Type Heat Exchanger. U.S. Patent 6684943B2, 3 February 2004.
10. Dobbs, G.M.; Benoit, J.T.; Lemcoff, N.O. Droplet Actuator Devices and Methods. WO2010/002957 A2, 1 July 2010.
11. Ehrenberg, S.G.; Huynh, H.; Johnson, B. Enhanced HVAC System and Method. U.S. Patent 8470071, 25 June 2013.
12. Dean, F.J.; Kadylak, D.E.; Huizing, R.N.; Balanko, J.B.; Mullen, C.W. Counter-Flow Energy Recovery Ventilator (ERV) Core. WO2013/091099 A1, 27 June 2013.
13. Alberti, G.; Narducci, R.; Di Vona, M.L.; Giancola, S. More on Nafion Conductivity Decay at Temperatures Higher than 80 °C: Preparation and First Characterization of In-Plane Oriented Layered Morphologies. *Ind. Eng. Chem. Res.* **2013**, *52*, 10418–10424. [CrossRef]
14. Alberti, G.; Di Vona, M.L.; Narducci, R. New results on the visco-elastic behaviour of ionomer membranes and relations between T-RH plots and proton conductivity decay of Nafion (R) 117 in the range 50–140 °C. *Int. J. Hydrogen Energy* **2012**, *37*, 6302–6307. [CrossRef]
15. Alberti, G.; Narducci, R.; Di Vona, M.L. *Solid State Proton Conductors: Properties and Applications in Fuel Cells*; Knauth, P., Di Vona, M.L., Eds.; Wiley: Hoboken, NJ, USA, 2012.
16. Mauritz, K.A.; Moore, R.B. State of understanding of Nafion. *Chem. Rev.* **2004**, *104*, 4535–4585. [CrossRef] [PubMed]
17. Roberti, E.; Carlotti, G.; Cinelli, S.; Onori, G.; Donnadio, A.; Narducci, R.; Casciola, M.; Sganappa, M. Measurement of the Young's modulus of Nafion membranes by Brillouin light scattering. *J. Power Sources* **2010**, *195*, 7761–7764. [CrossRef]
18. Yang, H.R.; Zhang, J.; Li, J.L.; Jiang, S.P.; Forsyth, M.; Zhu, H.J. Proton Transport in Hierarchical-Structured Nafion Membranes: A NMR Study. *J. Phys. Chem. Lett.* **2017**, *8*, 3624–3629. [CrossRef] [PubMed]
19. Buck, R.C.; Franklin, J.; Berger, U.; Conder, J.M.; Cousins, I.T.; Voogt, P.D.; Jensen, A.A.; Kannan, K.; Mabury, S.A.; Leeuwen, S.P.V. Perfluoroalkyl and polyfluoroalkyl substances in the environment: terminology, classification, and origins. *Integr. Environ. Assess. Manag.* **2011**, *7*, 513–541. [CrossRef] [PubMed]
20. *Identification of Substances of Very High Concern (SVHC) under the 'Equivalent Level of Concern' Route (REACH Article 57(f))—Neurotoxicants and Immunotoxicants as Examples*; EUR Scientific and Technical Research Reports; Publications Office of the European Union: Rue Mercier, Luxembourg, 2015. [CrossRef]
21. Alberti, G.; Narducci, R.; Sganappa, M. Effects of hydrothermal/thermal treatments on the water-uptake of Nafion membranes and relations with changes of conformation, counter-elastic force and tensile modulus of the matrix. *J. Power Sources* **2008**, *178*, 575–583. [CrossRef]
22. Alberti, G.; Narducci, R.; Di Vona, M.L.; Giancola, S. Annealing of Nafion 1100 in the Presence of an Annealing Agent: APowerful Method for Increasing Ionomer Working Temperature in PEMFCs. *Fuel Cells* **2013**, *13*, 42–47. [CrossRef]
23. Potreck, J. Membranes for Flue Gas Treatment. Ph.D. Thesis, University of Twente, Enschede, The Netherlands, 2009.

24. Sgreccia, E.; Chailan, J.F.; Khadhraoui, M.; Di Vona, M.L.; Knauth, P. Mechanical properties of proton-conducting sulfonated aromatic polymer membranes: Stress-strain tests and dynamical analysis. *J. Power Sources* **2010**, *195*, 7770–7775. [CrossRef]
25. Barbieri, G.; Brunetti, A.; Di Vona, M.L.; Sgreccia, E.; Knauth, P.; Hou, H.Y.; Hempelmann, R.; Arena, F.; Beretta, L.D.; Bauer, B.; et al. LoLiPEM: Long life proton exchange membrane fuel cells. *Int. J. Hydrogen Energy* **2016**, *41*, 1921–1934. [CrossRef]
26. Di Vona, M.L.; Sgreccia, E.; Licoccia, S.; Alberti, G.; Tortet, L.; Knauth, P. Analysis of Temperature-Promoted and Solvent-Assisted Cross-Linking in Sulfonated Poly(ether ether ketone) (SPEEK) Proton-Conducting Membranes. *J. Phys. Chem. B* **2009**, *113*, 7505–7512. [CrossRef] [PubMed]
27. Maranesi, B.; Hou, H.; Polini, R.; Sgreccia, E.; Alberti, G.; Narducci, R.; Knauth, P.; Di Vona, M.L. Cross-Linking of Sulfonated Poly(ether ether ketone) by Thermal Treatment: How Does the Reaction Occur? *Fuel Cells* **2013**, *13*, 107–117. [CrossRef]
28. Rammler, D.H.; Zaffaroni, A. Biological implications of DMSO based on a review of its chemical properties. *Ann. N. Y. Acad. Sci.* **1967**, *141*, 13–23. [CrossRef] [PubMed]
29. Di Vona, M.L.; Pasquini, L.; Narducci, R.; Pelzer, K.; Donnadio, A.; Casciola, M.; Knauth, P. Cross-linked sulfonated aromatic ionomers via SO_2 bridges: Conductivity properties. *J. Power Sources* **2013**, *243*, 488–493. [CrossRef]
30. Kerres, J.A. Blended and cross-linked ionomer membranes for application in membrane fuel cells. *Fuel Cells* **2005**, *5*, 230–247. [CrossRef]
31. Kerres, J.A.; Xing, D.M.; Schonberger, F. Comparative investigation of novel PBI blend ionomer membranes from nonfluorinated and partially fluorinated poly arylene ethers. *J. Polym. Sci. Part B Polym. Phys.* **2006**, *44*, 2311–2326. [CrossRef]
32. Bhattacharya, A.; Rawlins, J.W.; Ray, P. *Polymer Grafting and Crosslinking*; Wiley: Hoboken, NJ, USA, 2008.
33. Huang, R.Y.M.; Shao, P.H.; Burns, C.M.; Feng, X. Sulfonation of poly(ether ether ketone)(PEEK): Kinetic study and characterization. *J. Appl. Polym. Sci.* **2001**, *82*, 2651–2660. [CrossRef]
34. Shibuya, N.; Porter, R.S. Kinetics of peek sulfonation in concentrated sulfuric-acid. *Macromolecules* **1992**, *25*, 6495–6499. [CrossRef]
35. Deluca, N.W.; Elabd, Y.A. Polymer electrolyte membranes for the direct methanol fuel cell: A review. *J. Polym. Sci. Part B Polym. Phys.* **2006**, *44*, 2201–2225. [CrossRef]
36. Jiang, R.C.; Kunz, H.R.; Fenton, J.M. Investigation of membrane property and fuel cell behavior with sulfonated poly(ether ether ketone) electrolyte: Temperature and relative humidity effects. *J. Power Sources* **2005**, *150*, 120–128. [CrossRef]
37. Astill, T.; Xie, Z.; Shi, Z.Q.; Navessin, T.; Holdcroft, S. Factors Influencing Electrochemical Properties and Performance of Hydrocarbon-Based Electrolyte PEMFC Catalyst Layers. *J. Electrochem. Soc.* **2009**, *156*, B499–B508. [CrossRef]
38. Chen, J.H.; Li, D.R.; Koshikawa, H.; Asano, M.; Maekawa, Y. Crosslinking and grafting of polyetheretherketone film by radiation techniques for application in fuel cells. *J. Membr. Sci.* **2010**, *362*, 488–494. [CrossRef]
39. Fontananova, E.; Brunetti, A.; Trotta, F.; Biasizzo, M.; Drioli, E.; Barbieri, G. Stabilization of Sulfonated Aromatic Polymer (SAP) Membranes Based on SPEEK-WC for PEMFCs. *Fuel Cells* **2013**, *13*, 86–97. [CrossRef]
40. Di Vona, M.L.; Sgreccia, E.; Narducci, R.; Pasquini, L.; Hou, H.; Knauth, P. Stabilized sulfonated aromatic polymers by *in situ* solvothermal cross-linking. *Front. Energy Res.* **2014**, *2*, 39. [CrossRef]
41. Narducci, R.; Di Vona, M.L.; Knauth, P. Cation-conducting ionomers made by ion exchange of sulfonated poly-ether-ether-ketone: Hydration, mechanical and thermal properties and ionic conductivity. *J. Membr. Sci.* **2014**, *465*, 185–192. [CrossRef]
42. Al Lafi, A.G.; Hay, J.N. State of the water in crosslinked sulfonated poly(ether ether ketone). *J. Appl. Polym. Sci.* **2013**, *128*, 3000–3009. [CrossRef]
43. Arena, F.; Mitzel, J.; Hempelmann, R. Permeability and Diffusivity Measurements on Polymer Electrolyte Membranes. *Fuel Cells* **2013**, *13*, 58–64. [CrossRef]
44. Molla, S.; Compan, V. Polymer blends of SPEEK for DMFC application at intermediate temperatures. *Int. J. Hydrogen Energy* **2014**, *39*, 5121–5136. [CrossRef]

45. Zhao, Y.Y.; Tsuchida, E.; Choe, Y.K.; Ikeshoji, T.; Barique, M.A.; Ohira, A. Ab initio studies on the proton dissociation and infrared spectra of sulfonated poly(ether ether ketone) (SPEEK) membranes. *Phys. Chem. Chem. Phys.* **2014**, *16*, 1041–1049. [CrossRef] [PubMed]
46. Han, S.; Zhang, M.S.; Shin, J.; Lee, Y.S. A Convenient Crosslinking Method for Sulfonated Poly(ether ether ketone) Membranes via Friedel-Crafts Reaction Using 1,6-Dibromohexane and Aluminum Trichloride. *J. Appl. Polym. Sci.* **2014**, *131*, 40695. [CrossRef]
47. Zaidi, S.M.J.; Mikhailenko, S.D.; Robertson, G.P.; Guiver, M.D.; Kaliaguine, S. Proton conducting composite membranes from polyether ether ketone and heteropolyacids for fuel cell applications. *J. Membr. Sci.* **2000**, *173*, 17–34. [CrossRef]
48. Alberti, G.; Narducci, R. Evolution of Permanent Deformations (or Memory) in Nafion 117 Membranes with Changes in Temperature, Relative Humidity and Time, and Its Importance in the Development of Medium Temperature PEMFCs. *Fuel Cells* **2009**, *9*, 410–420. [CrossRef]
49. Alberti, G.; Narducci, R.; Di Vona, M.L.; Giancola, S. Preparation and Nc/T plots of un-crystallized Nafion 1100 and semi-crystalline Nafion 1000. *Int. J. Hydrogen Energy* **2017**, *42*, 15908–15912. [CrossRef]
50. Di Vona, M.L.; Baldinelli, G.; Marrocchi, A.; Narducci, R. Scambiatori di Calore Entalpici a Membrane di Tipo Polimerico Aromatic Solfonato e Procedimento per la Preparazione di Dette Membrane. Domanda numero: 102016000112268. 8 November 2016. (In Italian)
51. Narducci, R. Ion Conducting Membranes for Fuel Cell. Ph.D. Thesis, University of Rome Tor Vergata, Aix Marseille, Rome, 2014.
52. Hou, H.Y.; Maranesi, B.; Chailan, J.F.; Khadhraoui, M.; Polini, R.; Di Vona, M.L.; Knauth, P. Crosslinked SPEEK membranes: Mechanical, thermal, and hydrothermal properties. *J. Mater. Res.* **2012**, *27*, 1950–1957. [CrossRef]
53. Knauth, P.; Sgreccia, E.; Di Vona, M.L. Chemomechanics of acidic ionomers: Hydration isotherms and physical model. *J. Power Sources* **2014**, *267*, 692–699. [CrossRef]

 © 2018 by the authors. Licensee MDPI, Basel, Switzerland. This article is an open access article distributed under the terms and conditions of the Creative Commons Attribution (CC BY) license (http://creativecommons.org/licenses/by/4.0/).

Article

Waterborne Acrylate-Based Hybrid Coatings with Enhanced Resistance Properties on Stone Surfaces

Francesca Sbardella [1,2,*], Lucilla Pronti [3], Maria Laura Santarelli [1], José Marìa Asua Gonzàlez [2] and Maria Paola Bracciale [1,*]

1 Department of Chemical Engineering Materials Environment (DICMA) and Research Center in Science and Technology for the Preservation of Historical-architectural Heritage (CISTeC), Sapienza University of Rome, Rome 00184, Italy; marialaura.santarelli@uniroma1.it
2 Department of Applied Chemistry, University of the Basque Country, Donostia-San Sebastián 20018, Spain; jm.asua@ehu.es
3 Department of Basic and Applied Sciences for Engineering (SBAI), Sapienza University of Rome, Rome 00161, Italy; lucilla.pronti@uniroma1.it
* Correspondence: francesca.sbardella@uniroma1.it (F.S.); mariapaola.bracciale@uniroma1.it (M.P.B.)

Received: 10 July 2018; Accepted: 11 August 2018; Published: 15 August 2018

Abstract: The application of coating polymers to building materials is a simple and cheap way to preserve and protect surfaces from weathering phenomena. Due to its environmentally friendly character, waterborne coating is the most popular type of coating, and improving its performance is an important key of research. The study presents the results regarding the mechanical and photo-oxidation resistance of some water-based acrylic coatings containing SiO_2 nanoparticles obtained by batch miniemulsion polymerization. Coating materials have been characterized in terms of hydrophobic/hydrophilic behavior, mechanical resistance and surface morphology by means of water-contact angle, and scrub resistance and atomic force microscopy (AFM) measurements depending on silica-nanoparticle content. Moreover, accelerated weathering tests were performed to estimate the photo-oxidation resistance of the coatings. The chemical and color changes were assessed by Fourier-transform infrared spectroscopy (FTIR) and colorimetric measurements. Furthermore, the nanofilled coatings were applied on two different calcareous lithotypes (Lecce stone and Carrara Marble). Its properties, such as capillary water absorption and color modification, before and after accelerated aging tests, were assessed. The properties acquired by the addition of silica nanoparticles in the acrylic matrix can ensure good protection against weathering of stone-based materials.

Keywords: waterborne coatings; batch miniemulsion; weathering; stone preservation

1. Introduction

One of the major problems of building materials exposed to the outdoors are the environmental conditions that seriously affect their durability. Among abiotic factors, water penetration is considered a major liability, particularly for porous materials. A common practice to preserve building materials from weathering is to use water-repellent surfaces [1,2] due to their ability to control the transport of different fluids between the surface and the interior, besides having other useful properties, such as antibiofouling, antisticking, anticorrosion, stain resistance, and self-cleanability [3–7]. The most effective and cheap way to provide these properties is the application of polymer-coating compositions obtained from various monomers, like acrylics, fluorinated, and silicon-based materials [8,9]. Particularly, acrylic and methacrylic monomers are widely used in this way for the protection of walls, façades, consolidation of monuments, and cultural heritage sites. Some important parameters of this class of materials are transparency, water repellence, and being lightweight. In addition, to provide high durability for façades and floors, these surface treatments must have high adhesive power and good mechanical properties (scratch and wear resistance). However, serious durability issues arise when such acrylic systems are used on structures located outdoors. Moreover, protectives based on acrylic resins exhibit a scarce adhesion to porous substrates and provide insufficient water drainage and vapor permeability from the coated surface [1,10,11]. The treatments applied on building materials must be "breathable" to avoid physical phenomena, such as freezing/thawing cycles and salt crystallization, causing fast deterioration of stone materials [12,13]. Indeed, the stronger the hydrophobicity of the conservation material, the larger the stress between hydrophilic/hydrophobic interfaces is, and much more likely that detaching and peeling will occur [14].

Furthermore, the emerging demand of modern society to reduce the emission of volatile organic compounds (VOC) led the interest of the scientific community to develop a new type of coatings: waterborne coatings [15] that use water as a medium to disperse a resin, thus making these coatings ecofriendly and easy to apply. These coatings are environmentally friendly, as American and European regulations require waterborne coatings to have a VOC content of less than 3.5 pounds per gallon of water. In addition, due to the reduced VOC emissions during application, waterborne coatings reduce the risk of fires, are easier to clean up (creating fewer dangerous residues), and result in reduced worker exposure to organic vapors [16]. However, this class of coatings is usually weaker in mechanical performance and water resistance compared with their solvent-borne counterparts. Moreover, although waterborne coatings are easy to apply to any type of surface, the contact angle of the film rarely exhibits hydrophobic value. The key parameters of hydrophobic surface are the low surface energy of the material and the texture of the geometrical micro- or nanostructure of the surface [17–19]. In the last two decades, the application of nanosized inorganic particles, incorporated or in situ formed inside a polymeric matrix [10,20], have been largely exploited in order to compensate for weaknesses and to develop a nanometric structure that promotes the development of waterborne coatings. Inorganic nanoparticles can impart the polymeric matrix with water-repellent, mechanical, thermal, electrical, optical, or adhesive properties, as well as add new functionalities useful for tailored applications [21–25].

Among the numerous inorganic nanoparticles used in polymeric coating, nanosized silica is the most common used for the enhancement of mechanical and thermal properties, in terms of mechanical strength, modulus, and thermal stability, and also for the higher water resistance of water-based nanocomposite coatings [25–31]. However, the mechanical and thermal properties are related to the amount of added nanoparticles and to the type of polymerization process. Indeed, in some cases, the glass-transition temperature (T_g) and the elastic modulus increased with increasing silica content [32]. This behavior can be due to the reduction of polymer-chain mobility by the inclusion of rigid nanoparticles. Conversely, in other cases, high amounts of nanosilica reduced the T_g and the temperature for maximum mechanical damping. Furthermore, high amount of silica nanoparticles lead to a more brittle behavior [32–36].

Because of the intrinsic hydrophilicity of inorganic nanofillers, they can be directly incorporated into waterborne resins consisting of polymer latex and aqueous-polymer dispersion. However, when the aqueous nanocomposite resins are drying, the compatibility between the polymer chains and the inorganic nanofiller dominates the dispersion state of the nanofillers. Poor compatibility generally leads to aggregates of such nanofillers during drying [37].

Organic-inorganic polymer hybrids, based on combinations of polymers with metals, ceramics, or both, have been prepared by several synthesis methodologies [29,38]. Considering the wide variety in terms of properties of these two materials, a crucial point is to get a product as much homogeneous as possible. To get homogeneous systems it is important that the polymer and the inorganic components show a high compatibility, which is usually achieved by chemical interaction, for example, Van-der-Waals forces, hydrogen bridges, and coordinative or covalent bonds. Therefore, either the polymer chain provides sufficient compatibility with the often-hydrophilic inorganic species or the inorganic components are adapted to the polymer chain [39].

Recently, miniemulsion polymerization has turned out to be an attractive way to obtain hybrid nanocomposite particles, especially when the synthesis of more complex particles is involved (core-shell, raspberry-like, etc.) As the submicron monomer droplets become the dominant site for particle nucleation, these allow the production of nanocomposite particles [26,40].

In this work, waterborne nanostructured hybrid silica/polyacrylate coatings obtained by batch miniemulsion polymerization have been synthetized and characterized by means of atomic force microscopy (AFM), static water-contact angle and scrub-resistance measurements. Several coatings with different silica loading (2, 5, and 10 wbm%) were artificially aged in order to evaluate their stability to photo-oxidative weathering. Fourier transform infrared (ATR-FTIR) and colorimetric measurements were performed on the aged samples in order to elucidate the chemical and aesthetic changes produced by photo-oxidation. Capillary water absorption tests were also performed on two carbonatic stones treated with the different latexes in order to investigate the water-repellency capability of the synthetized coatings.

2. Materials and Methods

2.1. Materials

The 30 wt.% water suspension of colloidal silica with commercial name of LUDOX AS-30 (size 12 nm; SSA 220 $m^2 \cdot g^{-1}$) was supplied from Sigma-Aldrich (Madrid, Spain). n-butyl acrylate (nBA, Quimidroga, Barcelona, Spain), methyl methacrylate (MMA, Quimidroga), acrylamide (AM, Sigma-Aldrich), acrylic acid (AA, Sigma-Aldrich), stearylacrylate (SA, 97 wt.%, Sigma-Aldrich), alkyldiphenyloxide disulfonate (Dowfax 2A1, 45 wt.%, Dow Chemicals, Madrid, Spain), potassium persulfate (KPS, Sigma-Aldrich), and formic acid were used without any further purification. Deionized water (Milli-Q quality, MilliPore, Madrid, Spain) was used in all polymerization recipes. Hydroquinone (Sigma-Aldrich) was used to quench the reaction samples withdrawn at representative reaction times to monitor the progress of the process.

Two carbonatic lithotypes, Lecce stone (known as "Pietra solare" by Apulia, Italy) and Carrara marble (by Tuscany, Italy), were selected as substrates. Although these stones are mainly composed of calcite, they possess different porosity: Lecce stone—38.8%, Carrara marble—0.4%. Stone samples were obtained as $5 \times 5 \times 5$ cm^3 cubes.

2.2. Preparation and Application of the Latexes

Acrylic–inorganic composite latexes were synthesized via a miniemulsion polymerization by a two-step process according to the procedure already described elsewhere [41].

Initially, the oil phase was prepared by dissolving BA, MMA, AM, and AA (47, 47, 1, and 1 wt.%) with a 4 wbm% of a costabilizer (SA) under stirring for 15 min. At the same time, the aqueous phase is obtained mixing the anionic emulsifier (Dowfax 2A1, 1 or 2 wbm%), the iniziator (KPS, 2 wbm%), LUDOX AS-30 (2 or 5 or 10 wbm%), and water under stirring for 15 min. The pH was kept at 3.5 with formic acid. Then, the oil and the aqueous phase were brought together and mixed for 15 min. The final dispersion was sonified for 10 min at 70% of amplitude. During sonication, the flask was immersed in an icewater bath to avoid overheating.

Batch miniemulsion polymerizations were carried out in a 500-mL jacketed reactor equipped with a reflux condenser, stirrer, sampling device, and nitrogen inlet. The temperature was fixed at 70 °C for 3 h. Table 1 summarized the different formulations for each latex, to whom a name was assigned that described: the first letter of the silica used, i.e., "L" is for LUDOX 30; the first number referring to the percentage of the silica added (2, 5, or 10 wbm%), and the last number for the percentage of the surfactant used (1 or 2 wbm%). Furthermore, the conversion grade, the particle size, and Polydispersity Index (PDI) were reported. The particles size and the conversion grade were measured by Dynamic Light Scattering (DLS, Zetasizer Nano Series, Malvern Instrument, Malvern, UK) and gravimetric analysis, respectively. The reported average particle size (droplet size) values represent an average of 3 repeated measurements. The PDI that describes the degree of "nonuniformity" of a distribution (for a perfectly uniform sample, the PDI would be 0.0) was also measured. After 3 h, the polymerization was almost complete and was short-stopped by an aqueous solution of 0.1 wt.% of hydroquinone.

Before the application of any protective coatings, the stone specimens were washed with deionized water, dried in an oven for 7 days at 60 °C (±5 °C), then stored in a dry atmosphere, and weighed until constant mass (±0.1%) was reached according to UNI 10921:2001 [42].

The different latex dispersions were applied by brushing directly on the apical surface of the samples until apparent refusal (i.e., when the stone surface remained wet for 1 min) [43]. Three samples for each stone and for each treatment were prepared. After each treatment, the samples were weighed and then kept at 23 °C (±5 °C) until constant weight.

2.3. Characterization

Thin polyacrylate/silica hybrid films (≈13 µm) were applied on glass substrates with a film applicator (Octoplex film applicator, TQC B.V., Seregno, Italy) in order to have a homogeneous surface. The films were dried overnight at 23 °C and 55% relative humidity. On these samples, hydrophobicity and resistance to wet abrasion measurements were performed. Static water-contact angle measurements were carried out according to the standard sessile-drop method by using a Data Physics OCA 20-model goniometer (DataPhysics Instruments GmbH, Filderstadt, Germany). 5-µL deionized water was placed on the films and an average of minimum 10 measurements taken from different positions on the surface were done. Scrub-resistance tests were performed according to the ASTM standard [44]. AFM was also used to analyze the morphology of the films cast on glass substrates. A commercial AFM setup (Solver, NT-MDT, Moscow, Russia) equipped with standard silicon tips (NSC16, Mikromasch, Wetzlar, Germany) was used to collect micrographs of the sample surfaces in tapping mode in air and at room conditions. The root mean square roughness (R_q) was calculated on 19 × 19 µm^2 areas. An average of minimum 3 measurements taken from different locations on the surface were done.

Furthermore, accelerated aging tests by UV irradiation were carried out in accordance with the ISO 16474-3:2013 standard [45] on films and on coated stone samples. The climatic test chamber (Angelantoni Industrie S.r.l., Massa Martana, Italy) was equipped with medium-pressure ultraviolet lamps of mercury; these lamps emit in the full spectrum of ultraviolet (UVA, UVB, and UVC) with maximum emission peak in the UVA range at 366 nm. The total radiation that reached all the samples was 5.2 W·m^{-2} as measured by means of radiometer. The total exposure consisted of 62 cycles and parameters of 1 cycle were as follows: 45 min UV radiation on and dry, 45 min UV off and water spraying, 55 min UV on and dry, and 15 min UV off and water spraying. The total duration of the experiment simulated a normal natural aging of 130 days, with maximum solar exposure, taking Europe as reference for the calculations. After each cycle, the samples were rotated horizontally in order to avoid concentration of radiation in one place.

Colorimetric measurements were performed on film samples to verify color modification due to light exposition. The 3 chromatic CIELab coordinates were calculated starting from diffuse-reflectance spectra acquired in the UV–vis spectral range (300–800 nm) with an AvaSpec-2048 spectrophotometer (Avantes, Apeldoorn, The Netherlands) equipped with bifurcated fibers that collect the reflected light with a 45°/0° geometry. A Spectralon standard (Labshere SRS-99-010, 99% reflectance) was taken as reference for the reflectance spectra. In order to quantify the color differences, the notion of a just-noticeable difference (JND = 3) in stimuli has been used [46]. Furthermore, according to the literature, $\Delta E < 5$ was considered as corresponding to a not-significant variation [47]. The measurements were made before and after accelerated exposure, always on the same 8 positions on the film, with the aid of a locating mask, and the arithmetic mean was calculated. During color measurements, the Spectralon standard was placed below each film cast on glass substrates.

Infrared spectroscopy was used to explain the changes in coating films caused by UV degradation. The chemical changes were detected with a Fourier-transform infrared spectrometer (FTIR, Vertex 70,

Bruker Optik GmbH, Ettlingen, Germany) equipped with a single-reflection Diamond ATR cell, a standard MIR source (HeNe), and a room temperature DTGS detector. The ATR-FTIR spectra were recorded with 256 scans in the mid-infrared range (400–4000 cm^{-1}) at a resolution of 4 cm^{-1}.

The water capillarity absorption experiments were performed following the UNI EN 15801:2010 [48]. The tests were carried out on 3 samples for each coating formulation, before and after accelerated aging tests. The absorption coefficient (AC, kg/m^2·h$^{0.5}$), and the relative capillarity index (RCI) [43,49] were calculated. Furthermore, in order to test the hydrophobicity of the stone surface after each coating application, the water-contact angles were measured as reported above.

3. Results and Discussion

As shown in Table 1, a nearly total conversion of the monomers into polymers (81%–100%) was achieved. Furthermore, DLS measurements indicate that the particle sizes are in a range between 115 to 250 nm; finally, a very uniform dispersion in the latexes is evident, being the PDI (i.e., the square of the ratio of the standard deviation to the mean diameter size) values very close to 0.

Table 1. The latexes synthesized by miniemulsion polymerization with acrylates and LUDOX 30.

Sample	Composition (wt.%)	LUDOX AS-30 [2] (wbm%)	Dowfax 2A1 [3] (wbm%)	Conversion Grade (%)	D (nm)	PDI
NoSiO$_2$	–	0	1	81.2	142	0.022
L2_1	BA: 47	2	1	100	199.7	0.017
L2_2	MMA: 47	2	2	96.3	156.1	0.145
L5_1	AM: 1	5	1	91.2	212.4	0.123
L5_2	AA: 1	5	2	81.5	115.4	0.029
L10_1	SA [1] : 4	10	1	97	250	0.101
L10_2	KPS [1] : 2	10	2	98.9	122.8	0.044

[1] weight based on monomers (wbm%); [2] colloidal silica, 30 wt.% water suspension; [3] alkyldiphenyloxide disulfonate. The values of the conversion, particle size (D), and Polydispersity Index (PDI) are referred to 180 min, i.e., the end of the polymerization process.

Surface hydrophobicity of the coatings is mandatory for protection applications due to their ability to form a protective layer able to control the transport of different fluids between the surface and the bulk interior. Essential requirements for a hydrophobic surface are a low surface energy of material and a high roughness with micro- or nanostructured surface architecture [4].

As previous studies have pointed out, a liquid either follows the surface (Wenzel scenario) [17] or it leaves air inside the texture (Cassie-Baxter) [18]. In the first case, surface roughness r (the ratio between the true surface area over the apparent one) is the key factor that controls contact angle; in the second case, a liquid is sitting upon a patchwork of air and solid surface. In the latter case, the contact angle is the average between the angle on the solid and the one on the air.

Figure 1a shows that the water-contact angle on polyacrylic coating (NoSiO$_2$) was 55° ± 0.2°. The incorporation of SiO$_2$ nanoparticles lead to contact-angle values generally higher than 80°, due to the increase in surface roughness (Figure 1b). Furthermore, the high level of hydrophobicity of the coatings with 1 wbm% of surfactant could be attributed to the crosslinking in the coating and limited surfactant exudation due to polymer–surfactant complex formation through adsorption of surfactant onto hydrophobic polymer segments [50]. Indeed, the enhanced wettability was recognized to result from migration of the hydrophilic segment of the surfactant to the latex film surface [50].

The dispersion grade of the silica nanoparticles within the film is another important factor affecting the hydrophobic properties of the surfaces. AFM images of 19 µm × 19 µm performed on all the nanocomposite coatings cast on glass substrates are shown in Figure 2. In the micrographs, it is observed that all the coatings exhibit a pillarlike surface fairly rough with a better dispersion of the SiO$_2$ nanoparticles in films with 1 wbm% (Figure 2b,d,f) than in those with 2 wbm% (Figure 2c,e,g) of surfactant. However, by increasing the amount of the silica, it is likely to find more spaced agglomerated particles of silica (Figure 2f) that lead to a decrease in the air pocket between the asperities.

The higher water-contact angles were indeed attained on samples having finer, more homogeneously spread, and high pillarlike structures [4]. The agglomeration phenomenon is pretty much evident in the latexes with 2 wbm% of Dowfax, where large uneven clusters of SiO_2 nanoparticles have settled. The standard deviation calculated from the R_q values (Figure 1b) confirms that a better homogeneous dispersion of silica nanoparticles can be achieved in the latexes with 1 wbm% of surfactant.

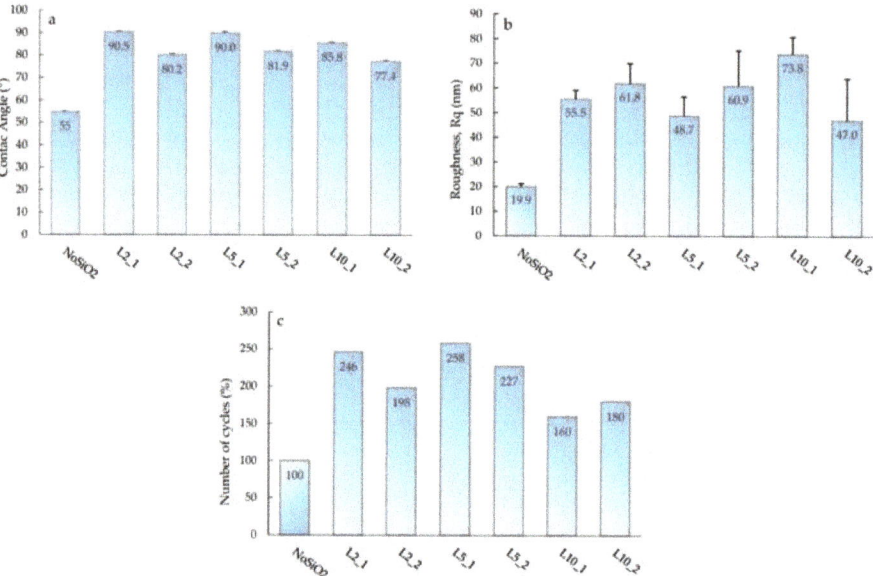

Figure 1. (**a**) Contact angle, (**b**) roughness (R_q) and (**c**) scrub resistance of the latex films synthesized with LUDOX 30.

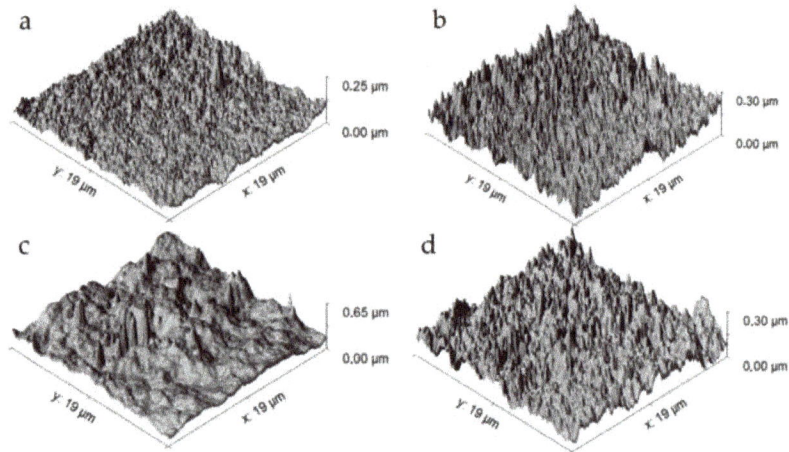

Figure 1. *Cont.*

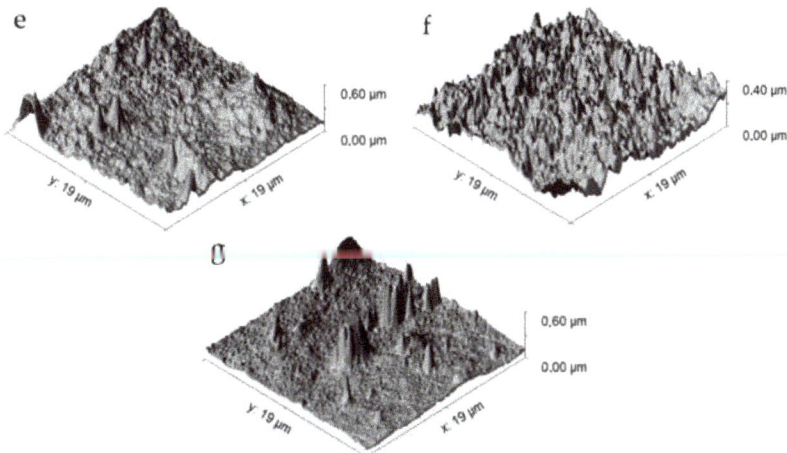

Figure 2. 3D atomic force microscopy (AFM) micrographs of hybrid coatings: (**a**) NoSiO$_2$; (**b**) L2_1; (**c**) L2_2; (**d**) L5_1; (**e**) (L5_2); (**f**) L10_1; and (**g**) L10_2.

When the roughness data and water-contact angles were correlated, large agglomerates of the silica nanoparticles were identified leading to a hierarchical micro/nanostructure on the surface that enhances the wettability of the surfaces as previously reported [51].

In order to be applied as interior-exterior finishes, it is essential to evaluate the capability of the investigated nanocomposites not to be easily scratched and damaged. To this end, washability was evaluated by determining the resistance of the film to wet erosion by visual assessment; this parameter is also referred to as "resistance to scrubbing" or "resistance to wet abrasion" and is determined with the number of back-and-forth strokes (cycles) required to remove the film. The results show (Figure 1c) that all the latexes containing the nanosilica have an increase in resistance to wet abrasion with respect to the one without silica (NoSiO$_2$). Probably, the lowest scrub-resistance values of the two latexes containing 10 wbm% of silica (L10_1 and L10_2) are due to a nonhomogeneous dispersion of the silica into the polymer matrix.

Finally, the optimal nanoparticle content for good dispersion, enhanced hydrophobicity, and mechanical resistance was found to be 5 wbm%.

Accelerated Aging

The chemical changes to the structure (bond scission/forming) of copolymers occurring upon UV/condensation-accelerated aging were monitored using FTIR-ATR and colorimetric analysis. Figure 3a shows the FTIR spectra of nonirradiated copolymer without and with 10 wbm% of silica. The spectrum of the neat polymer (Figure 3a, grey curve) exhibits several characteristic spectral bands, such as: −OH stretch (3444 cm^{-1}), C–H stretch (2956 to 2874 cm^{-1}), C=O stretch (1726 cm^{-1}), C–H bending (1386 and 1450 cm^{-1}), C–C–O stretch (1267 and 1236 cm^{-1}), C–O–C stretch (1160 and 1145 cm^{-1}), O=C–O stretch (1063 cm^{-1}), C–H bending in MMA (989 cm^{-1}), C–H bending in nBA (962 cm^{-1}), and vibrations of the side chains (842 and 753 cm^{-1}) [52]. After the addition of the inorganic components (Figure 3a, black curve), strong absorption bands appear at 1118 and 476 cm^{-1} due to Si–O–Si stretching and bending, respectively, and at 1068 cm^{-1} related to Si–O bending. As shown in Figure 3b, no band shifting was observed in the UV-irradiated sample with reference to the nonirradiated sample. Particularly, the most remarkable changes detected in FTIR spectra were the increase in the absorption of the hydroxyl region, between 3400 and 3200 cm^{-1}, due to oxidation reactions. Furthermore, relevant changes after aging are observed in the region of 1800–1600 cm^{-1}.

The absorption of carbonyl band (1726 cm^{-1}) shows evident broadening, with the appearance of a shoulder at about 1780 and 1640 cm^{-1}. Those effects may be attributed to a γ-lactone [43,53–55] and to terminal C=C double bonds, respectively, produced by chain scission increasing the terminal carbon–carbon unsaturations during aging [53,54].

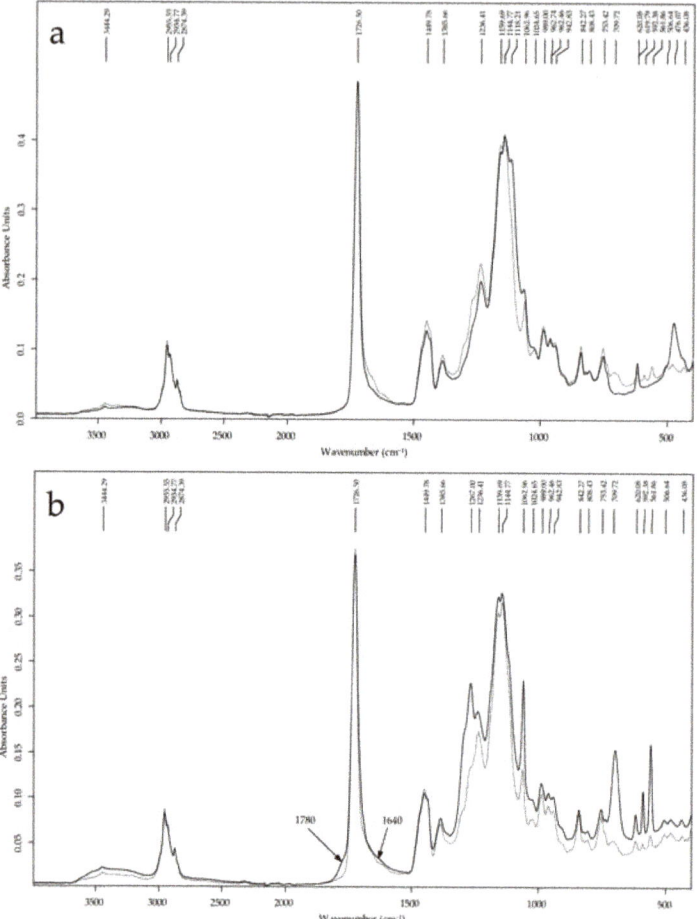

Figure 3. (a) Fourier-transform infrared spectroscopy (ATR-FTIR) spectra of latex films without (grey) and with 10 wt.% (black) of nanosilica before UV exposure; (b) ATR-FTIR spectra of the neat films without SiO$_2$ before (grey) after (black) UV exposure.

Furthermore, minor changes in band intensities were observed, evidencing chemical changes in the copolymer upon UV irradiation. The band underwent prominent changes is the carbonyl (C=O) stretching, which variation, as previously reported [52], was quantified after UV exposure. Indeed, considering the decrease of this band, the relative amount of the remaining functional group was determined by the ratio of IR absorbance at 1726 cm^{-1} after aging and the absorbance of the unexposed sample (Figure 4a) [55].

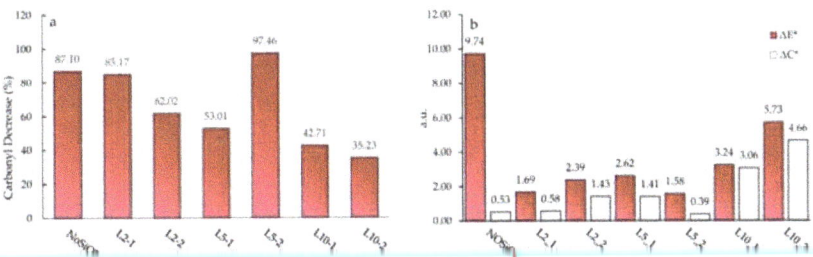

Figure 4. (a) Carbonyl decrease and (b) CIELaB determinations of the UV irradiated latexes.

The addition of nanoparticles changed the chemical structure of acrylic coatings modifying their resistances to UV-aging conditions. Figure 4a shows the calculation of the IR intensity changes of the C=O functional group for all the coatings analyzed. The presence of 5 wbm% of SiO_2 nanoparticles slowed the degradation of acrylic coating, as evidenced by the less-pronounced decrease of the carbonyl band in comparison with the neat coating. A higher content of inorganic nanoparticles is detrimental to the chemical stability of the films, probably due to the formation of large aggregates that inhibit the formation of a continuous cross-linked film. Furthermore, a general tendency in the photo-oxidation can be seen in coatings with 2 wbm% of surfactant. This phenomenon might be due to exudation and degradation of surfactant to the surface under light aging that produce hydrophilic and oxidized small molecules, which could promote radical and photochemical-degradation processes on the film surface, as previously reported [56]. The weathering of the polymers was also assessed by colorimetric changes in the total color differences (ΔE^*) and in the metric chroma-color difference (ΔC^*) using the following equations:

$$\Delta E^* = \sqrt{(\Delta L^*)^2 + (\Delta a^*)^2 + (\Delta b^*)^2} \qquad (1)$$

$$\Delta C^* = \sqrt{(\Delta a^*)^2 + (\Delta b^*)^2} \qquad (2)$$

where L^*, a^*, and b^* are the brightness, the red–green component, and the yellow–blue component, respectively [46]. The total color change parameter is important for aesthetic reasons, since a coating should not induce ΔE^* greater than 3 [46,47], in order to preserve the original color of surfaces. The ΔE^* and ΔC^* values after aging are reported in Figure 4b. The neat polymer, being subjected to depolymerization after photolytic scission of main chains, shows the largest photodegradation [43]. Indeed, the highest ΔE^* value due to an increase in the L^* coordinate up to the value of the Spectralon reference (Figure S1 and Table S1 in Supplementary Materials) could be a consequence of polymer loss that lead gradually to several uncoated areas [53]. No yellowing effect was detected, as proved by the negligible C variation. The addition of the inorganic nanoparticles, forming a 3D network that slows down UV photo-oxidation, is in agreement with the FTIR results: in this case, the color differences are more evident in the coatings with a larger amount of SiO_2 nanoparticles (10 wbm%). The latter exhibits shift towards yellow coordination with reduced lightness due to the occurrence of chain scission (Figure S1 and Table S1 in Supplementary Materials).

Another important factor to be considered is the behavior of the coatings to contrast the absorption of water. Indeed, water is one of the most important factors of deterioration in porous materials thanks to its capability to penetrate by capillary absorption. Furthermore, water is the main vehicle for the transport of gaseous pollutants that can attack the components of the matrix, and physical phenomena can be induced, such as freezing/thawing cycles and salt crystallization, causing fast deterioration of the stone materials.

To evaluate the usefulness of the coatings in decreasing water penetration, water capillary-absorption tests were carried out [48]. The measurements were performed on treated and untreated stone before and after accelerated-aging tests, simulating the coating performance under solar irradiation, which may lead to a decrease in water resistance due to chemical modifications in the polymer films. The absorption coefficient (AC), calculated as the slope of the linear part of the absorption curve in the first hour, the relative capillary index (RCI), obtained by the ratio of the water absorbed by the treated specimen and the water absorbed by the untreated one, and the contact-angle values measured on untreated and treated stones are reported in Table 2.

Table 2. Values of the contact angles, absorption coefficient (AC), and relative capillary index (RCI) of untreated and treated Lecce stone and Carrara marble before and after aging tests.

Stone	Coating	Contact Angle [1]	Before Aging		After Aging	
			AC [2]	RCI	AC [2]	RCI
Lecce stone	Untreated	0	108.5	1.00	108.5	1.00
	NoSiO$_2$	71.08 ± 1.11	36.1	0.80	38.5	0.84
	L2_1	93.95 ± 0.49	10.90	0.78	11.96	0.75
	L2_2	91.25 ± 0.67	13.38	0.85	9.87	0.78
	L5_1	92.69 ± 0.49	12.02	0.83	16.66	0.82
	L5_2	90.46 ± 0.76	14.06	0.87	10.21	0.79
	L10_1	89.77 ± 0.65	12.57	0.84	16.67	0.83
	L10_2	88.93 ± 1.32	13.65	0.85	10.65	0.78
Carrara marble	Untreated	54.08 ± 0.68	0.23	1.00	0.23	1.00
	NoSiO$_2$	76.85 ± 0.21	0.13	0.93	0.11	0.99
	L2_1	94.60 ± 0.28	0.08	0.51	0.10	0.37
	L2_2	91.16 ± 0.45	0.10	0.63	0.08	0.35
	L5_1	92.82 ± 0.48	0.10	0.94	0.06	0.55
	L5_2	92.33 ± 0.56	0.09	0.62	0.13	0.37
	L10_1	92.15 ± 0.88	0.10	0.92	0.07	0.55
	L10_2	88.28 ± 1.21	0.07	0.60	0.15	0.38

[1] Degree ± standard deviation; [2] kg/m$^2 \cdot$h$^{0.5}$.

The hydrophobic/hydrophilic behavior of the unaged stone, treated with the synthetized latexes, was firstly evaluated through static contact angle, evidencing high superficial hydrophobicity of the stone treated with nanosilica, in respect to neat latex. The results were in accordance with the former measurements performed on films casted on glass substrates.

All the treatments with polymer–nanoparticle coating affected the capillary absorption behavior compared to the untreated stones and reduced water absorption, as shown from the AC and RCI values. In detail, from the AC values, which depict the speed of the capillary rise at low times, it is clear how the absorption of water in the presence of the latexes is halved compared to the untreated reference for the latexes at 1 wbm% of surfactant. According to the literature, an effective hydrophobic treatment should reduce the capillary water-AC to 0.1 kg/m$^2 \cdot$h$^{0.5}$ a value small enough to provide sufficient protection against driving rain [57]. The AC values for the Carrara marble-treated stones are lower than 0.1 kg/m$^2 \cdot$h$^{0.5}$, meaning that the treatments have generally hydrophobic behavior, reducing the absorption of water by capillary effect. This hydrophobic behavior is also evident for Lecce-stone values, where the decreasing amount of water absorbed showed sharp reduction, even if the values of AC are higher than the limit proposed (0.1). This behavior could probably be due to the different porosity values of the stones and the evidently higher amount of water absorbed by Lecce stone.

After aging, the treated samples with latex without silica (NoSiO$_2$) showed an increase in the RCI value, revealing a decrease in the hydrophobic features of the organic polymer, probably due to loss of the polymer applied, as above reported for the colorimetric measurements. All the hybrid coatings are particularly stable on both the lithotypes, showing small variations of the RCI.

4. Conclusions

Hybrid waterborne coatings based on acrylate copolymer containing SiO_2 nanoparticles were synthesized by batch miniemulsion polymerization. The addition of nanosilica in the polymer coating increased surface roughness by the creation of nanoscale-structured surface architecture, as confirmed by AFM micrographs. This structure allowed a substantial increase in water-contact angles, creating a surface with good hydrophilic/hydrophobic balance. Furthermore, the addition of 5 wbm% of nanoparticles enhanced the mechanical scrub resistance and the stability of the polymer coating to photoaging thanks to the formation of homogenous organic-inorganic cross-linked structure. The accelerated photoaging of the coated stones confirms the positive effect of the inorganic nanoparticles in reducing capillary absorption, in particular at short times, and in enhancing coating stability. Therefore, the properties acquired by the addition of silica nanoparticles in the acrylic matrix can ensure good protection against the weathering of stone-based materials.

Supplementary Materials: The following are available online at http://www.mdpi.com/2079-6412/8/8/283/s1, Figure S1: CIELaB measurements of the UV irradiated latexes, Table S1: Values of the colorimetric measurements.

Author Contributions: Conceptualization, M.L.S. and J.M.A.G.; Funding Acquisition, M.L.S.; Investigation, F.S., L.P., and M.P.B.; Methodology, F.S.; Project Administration, M.L.S. and J.M.A.G.; Resources, M.L.S. and J.M.A.G.; Validation, M.P.B.; Visualization, F.S. and M.P.B.; Writing-Original Draft, F.S. and M.P.B.; Writing-Review & Editing, M.L.S. and J.M.A.G.

Funding: This research received no external funding.

Acknowledgments: The authors would like to thank Daniele Passeri (Department of Basic and Applied Sciences for Engineering (SBAI), Sapienza University of Rome) for help with AFM, and Maria Gabriella Santonicola (Department of Chemical Engineering Materials Environment-DICMA, Sapienza University of Rome) for her kind assistance in the contact-angle measurements on tilted plane. M.L.S. thanks Paul Smith and Walter Remo Caseri (ETH Zurich) for their personal support to the project.

Conflicts of Interest: The authors declare no conflict of interest.

References

1. Doehne, E.F.; Price, C.A. *Stone Conservation: An Overview of Current Research*; Getty Conservation Institute: Los Angeles, CA, USA, 2010; ISBN 1606060465.
2. Khallaf, M.K.; El-Midany, A.A.; El-Mofty, S.E. Influence of acrylic coatings on the interfacial, physical, and mechanical properties of stone-based monuments. *Prog. Org. Coat.* **2011**, *72*, 592–598. [CrossRef]
3. López, A.B.; De La Cal, J.C.; Asua, J.M. Highly Hydrophobic Coatings from Waterborne Latexes. *Langmuir* **2016**, *32*, 7459–7466. [CrossRef] [PubMed]
4. Kamegawa, T.; Irikawa, K.; Yamashita, H. Multifunctional surface designed by nanocomposite coating of polytetrafluoroethylene and TiO_2 photocatalyst: Self-cleaning and superhydrophobicity. *Sci. Rep.* **2017**, *7*, 13628. [CrossRef] [PubMed]
5. Wan, H.; Song, D.; Li, X.; Zhang, D.; Gao, J.; Du, C. Failure Mechanisms of the Coating/Metal Interface in Waterborne Coatings: The Effect of Bonding. *Materials* **2017**, *10*, 397. [CrossRef] [PubMed]
6. Wang, N.; Diao, X.; Zhang, J.; Kang, P. Corrosion Resistance of Waterborne Epoxy Coatings by Incorporation of Dopamine Treated Mesoporous-TiO_2 Particles. *Coatings* **2018**, *8*, 209. [CrossRef]
7. Raditoiu, V.; Raditoiu, A.; Raduly, M.; Amariutei, V.; Gifu, I.; Anastasescu, M. Photocatalytic Behavior of Water-Based Styrene-Acrylic Coatings Containing TiO_2 Sensitized with Metal-Phthalocyanine Tetracarboxylic Acids. *Coatings* **2017**, *7*, 229. [CrossRef]
8. Cappelletti, G.; Fermo, P.; Camiloni, M. Smart hybrid coatings for natural stones conservation. *Prog. Org. Coat.* **2015**, *78*, 511–516. [CrossRef]
9. Calia, A.; Colangiuli, D.; Lettieri, M.; Matera, L. A deep knowledge of the behaviour of multi-component products for stone protection by an integrated analysis approach. *Prog. Org. Coat.* **2013**, *76*, 893–899. [CrossRef]
10. Esposito Corcione, C.; De Simone, N.; Santarelli, M.L.; Frigione, M. Protective properties and durability characteristics of experimental and commercial organic coatings for the preservation of porous stone. *Prog. Org. Coat.* **2017**, *103*, 193–203. [CrossRef]

11. Pia, G.; Esposito Corcione, C.; Striani, R.; Casnedi, L.; Sanna, U. Coating's influence on water vapour permeability of porous stones typically used in cultural heritage of Mediterranean area: Experimental tests and model controlling procedure. *Prog. Org. Coat.* **2017**, *102*, 239–246. [CrossRef]
12. Tabasso, M.L. Acrylic Polymers for the Conservation of Stone: Advantages and Drawbacks. *APT Bull. J. Preserv. Technol.* **1995**, *26*, 17–21. [CrossRef]
13. Siegesmund, S.; Snethlage, R. *Stone in Architecture: Properties, Durability*; Springer Science & Business Media: Berlin, Germany, 2013; ISBN 3642451551.
14. Zhang, H.; Liu, Q.; Liu, T.; Zhang, B. The preservation damage of hydrophobic polymer coating materials in conservation of stone relics. *Prog. Org. Coat.* **2013**, *76*, 1127–1134. [CrossRef]
15. Paulis, M.; Asua, J.M. Knowledge-Based Production of Waterborne Hybrid Polymer Materials. *Macromol. React. Eng.* **2016**, *10*, 8–21. [CrossRef]
16. Judeinstein, P.; Sanchez, C. Hybrid organic/inorganic materials: A land of multidisciplinarity. *J. Mater. Chem.* **1996**, *6*, 511–525. [CrossRef]
17. Wenzel, R.N. Resistance of solid surfaces to wetting by water. *Ind. Eng. Chem.* **1936**, *28*, 988–994. [CrossRef]
18. Cassie, A.B.D.; Baxter, S. Wettability of porous surfaces. *Trans. Faraday Soc.* **1944**, *40*, 546–551. [CrossRef]
19. Feng, L.; Li, S.; Li, Y.; Li, H.; Zhang, L.; Zhai, J.; Song, Y.; Liu, B.; Jiang, L.; Zhu, D. Super-hydrophobic surfaces: From natural to artificial. *Adv. Mater.* **2002**, *14*, 1857–1860. [CrossRef]
20. Gagliardi, S.; Rondino, F.; D'erme, C.; Persia, F.; Menchini, F.; Santarelli, M.L.; Paulke, B.-R.; Enayati, L. Preparation and characterization of polymeric nanocomposite films for application as protective coatings. In Proceedings of the AIP Conference Proceedings, Rome, Italy, 20–23 September 2016; Volume 1873, pp. 20007–20009.
21. Kim, E.K.; Won, J.; Do, J.; Kim, S.D.; Kang, Y.S. Effects of silica nanoparticle and GPTMS addition on TEOS-based stone consolidants. *J. Cult. Herit.* **2009**, *10*, 214–221. [CrossRef]
22. Dei, L.; Salvadori, B. Nanotechnology in cultural heritage conservation: Nanometric slaked lime saves architectonic and artistic surfaces from decay. *J. Cult. Herit.* **2006**, *7*, 110–115. [CrossRef]
23. Xu, F.; Wang, C.; Li, D.; Wang, M.; Xu, F.; Deng, X. Preparation of modified epoxy-SiO_2 hybrid materials and their application in the stone protection. *Prog. Org. Coat.* **2015**, *81*, 58–65. [CrossRef]
24. Illescas, J.F.; Mosquera, M.J. Producing Surfactant-Synthesized Nanomaterials In Situ on a Building Substrate, without Volatile Organic Compounds. *ACS Appl. Mater. Interfaces* **2012**, *4*, 4259–4269. [CrossRef] [PubMed]
25. Kugler, S.; Kowalczyk, K.; Spychaj, T. Influence of dielectric nanoparticles addition on electroconductivity and other properties of carbon nanotubes-based acrylic coatings. *Prog. Org. Coat.* **2016**, *92*, 66–72. [CrossRef]
26. Zou, H.; Wu, S.; Shen, J. Polymer/silica nanocomposites: Preparation, characterization, properties and applications. *Chem. Rev.* **2008**, *108*, 3893–3957. [CrossRef] [PubMed]
27. Ma, J.-Z.; Hu, J.; Zhang, Z.-J. Polyacrylate/silica nanocomposite materials prepared by sol–gel process. *Eur. Polym. J.* **2007**, *43*, 4169–4177. [CrossRef]
28. Chau, J.L.H.; Hsieh, C.C.; Lin, Y.M.; Li, A.K. Preparation of transparent silica-PMMA nanocomposite hard coatings. *Prog. Org. Coat.* **2008**, *62*, 436–439. [CrossRef]
29. Ribeiro, T.; Baleizão, C.; Farinha, J.P.S. Functional films from silica/polymer nanoparticles. *Materials* **2014**, *7*, 3881–3900. [CrossRef] [PubMed]
30. Zhang, K.; Zheng, L.; Zhang, X.; Chen, X.; Yang, B. Silica-PMMA core-shell and hollow nanospheres. *Colloids Surf. A Physicochem. Eng. Asp.* **2006**, *277*, 145–150. [CrossRef]
31. Bao, Y.; Ma, J.; Zhang, X.; Shi, C. Recent advances in the modification of polyacrylate latexes. *J. Mater. Sci.* **2015**, *50*, 6839–6863. [CrossRef]
32. Romo-Uribe, A.; Arcos-Casarrubias, J.A.; Hernandez-Vargas, M.L.; Reyes-Mayer, A.; Aguilar-Franco, M.; Bagdhachi, J. Acrylate hybrid nanocomposite coatings based on SiO_2 nanoparticles by in-situ batch emulsion polymerization. *Prog. Org. Coat.* **2016**, *97*, 288–300. [CrossRef]
33. Mahaling, R.N.; Kumar, S.; Rath, T.; Das, C.K. Acrylic elastomer/filler nanocomposite: Effect of silica nanofiller on thermal, mechanical and interfacial adhesion. *Plast. Rubber Compos.* **2007**, *36*, 267–273. [CrossRef]
34. Bandyopadhyay, A.; Bhowmick, A.K.; De Sarkar, M. Synthesis and characterization of acrylic rubber/silica hybrid composites prepared by sol-gel technique. *J. Appl. Polym. Sci.* **2004**, *93*, 2579–2589. [CrossRef]
35. Arai, K.; Mizutani, T.; Kimura, Y.; Miyamoto, M. Unique structure and properties of inorganic-organic hybrid films prepared from acryl/silica nano-composite emulsions. *Prog. Org. Coat.* **2016**, *93*, 109–117. [CrossRef]

36. Ramos-Fernández, J.M.; Beleña, I.; Romero-Sánchez, M.D.; Fuensanta, M.; Guillem, C.; López-Buendía, Á.M. Study of the film formation and mechanical properties of the latexes obtained by miniemulsion co-polymerization of butyl acrylate, methyl acrylate and 3-methacryloxypropyltrimethoxysilane. *Prog. Org. Coat.* **2012**, *75*, 86–91. [CrossRef]
37. Zhou, S.; Wu, L. Transparent Organic-Inorganic Nanocomposite Coatings. In *Functional Polymer Coatings: Principles, Methods, and Applications*; Wu, L., Baghdachi, J., Eds.; Wiley: Hoboken, NJ, USA, 2015; pp. 1–71.
38. Krasia-Christoforou, T. Organic-inorganic polymer hybrids: Synthetic strategies and applications. In *Hybrid and Hierarchical Composite Materials*; Springer International Publishing: Cham, Switzerland, 2015; pp. 11–63. ISBN 9783319128689.
39. Kickelbick, G. Hybrid Materials—Past, Present and Future. *Hybrid Mater.* **2014**, *1*, 39–51. [CrossRef]
40. Bourgeat-Lami, E.; Lansalot, M. Organic/Inorganic Composite Latexes: The Marriage of Emulsion Polymerization and Inorganic Chemistry. In *Hybrid Latex Particles: Preparation with (Mini) Emulsion Polymerization*; Van Herk, A.M., Landfester, K., Eds.; Springer: Berlin/Heidelberg, Germany, 2010; Volume 233, pp. 53–123. ISBN 978-3-642-16060-8.
41. Hamzehlou, S.; Aguirre, M.; Leiza, J.R.; Asua, J.M. Dynamics of the Particle Morphology during the Synthesis of Waterborne Polymer-Inorganic Hybrids. *Macromolecules* **2017**, *50*, 7190–7201. [CrossRef]
42. *UNI 10921:2001 Cultural Heritage—Natural and Artificial Stones—Water Repellents—Application on Samples and Determination of their Properties in Laboratory*; UNI: Milan, Italy, 2001.
43. Bergamonti, L.; Bondioli, F.; Alfieri, I.; Alinovi, S.; Lorenzi, A.; Predieri, G.; Lottici, P.P. Weathering resistance of PMMA/SiO_2/ZrO_2 hybrid coatings for sandstone conservation. *Polym. Degrad. Stab.* **2018**, *147*, 274–283. [CrossRef]
44. *ASTM D2486-00 Standard Test Methods for Scrub Resistance of Wall Paints*; ASTM International: West Conshohocken, PA, USA, 2000.
45. *ISO 16474-3:2013 Paints and Varnishes—Methods of Exposure to Laboratory Light Sources—Part 3: Fluorescent UV Lamps 2013*; ISO: Geneva, Switzerland, 2013.
46. Sharma, G.; Bala, R. *Digital Color Imaging Handbook*, 1st ed.; Sharma, G., Ed.; Electrical Engineering & Applied Signal Processing Series; CRC Press: Boca Raton, FL, USA, 2002; ISBN 978-0-8493-0900-7.
47. Mahy, M.; Van Eycken, L.; Oosterlinck, A. Evaluation of Uniform Color Spaces Developed after the Adoption of CIELAB and CIELUV. *Color Res. Appl.* **1994**, *19*, 105–121. [CrossRef]
48. *UNI EN 15801:2010 Conservation of Cultural Heritage—Test Methods—Determination of Water Absorption by Capillarity*; UNI: Milan, Italy, 2010.
49. Washburn, E.W. The dynamics of capillary flow. *Phys. Rev.* **1921**, *17*, 273–283. [CrossRef]
50. Li, J.; Ecco, L.; Delmas, G.; Whitehouse, N.; Collins, P.; Deflorian, F.; Pan, J. In-Situ AFM and EIS Study of Waterborne Acrylic Latex Coatings for Corrosion Protection of Carbon Steel. *J. Electrochem. Soc.* **2014**, *162*, C55–C63. [CrossRef]
51. Vázquez-Velázquez, A.; Velasco-Soto, M.; Pérez-García, S.; Licea-Jiménez, L. Functionalization Effect on Polymer Nanocomposite Coatings Based on TiO_2–SiO_2 Nanoparticles with Superhydrophilic Properties. *Nanomaterials* **2018**, *8*, 369. [CrossRef] [PubMed]
52. Shanti, R.; Hadi, A.N.; Salim, Y.S.; Chee, S.Y.; Ramesh, S.; Ramesh, K. Degradation of ultra-high molecular weight poly(methyl methacrylate-co-butyl acrylate-co-acrylic acid) under ultra violet irradiation. *RSC Adv.* **2017**, *7*, 112–120. [CrossRef]
53. Melo, M.; Bracci, S.; Camaiti, M.; Chiantore, O.; Piacenti, F. Photodegradation of acrylic resins used in the conservation of stone. *Polym. Degrad. Stab.* **1999**, *66*, 23–30. [CrossRef]
54. Allen, N.S.; Regan, C.J.; McIntyre, R.; Johnson, B.W.; Dunk, W.A.E. The photooxidative degradation and stabilisation of water-borne acrylic coating systems. *Macromol. Symp.* **1997**, *115*, 1–26. [CrossRef]
55. Nguyen, T.V.; Nguyen Tri, P.; Nguyen, T.D.; El Aidani, R.; Trinh, V.T.; Decker, C. Accelerated degradation of water borne acrylic nanocomposites used in outdoor protective coatings. *Polym. Degrad. Stab.* **2016**, *128*, 65–76. [CrossRef]

56. Scalarone, D.; Lazzari, M.; Castelvetro, V.; Chiantore, O. Surface monitoring of surfactant phase separation and stability in waterborne acrylic coatings. *Chem. Mater.* **2007**, *19*, 6107–6113. [CrossRef]
57. Winkler, E.M. *Stone in Architecture*; Springer: Berlin/Heidelberg, Germany, 1997; ISBN 978-3-662-10072-1.

© 2018 by the authors. Licensee MDPI, Basel, Switzerland. This article is an open access article distributed under the terms and conditions of the Creative Commons Attribution (CC BY) license (http://creativecommons.org/licenses/by/4.0/).

Review

Novel Attribute of Organic–Inorganic Hybrid Coatings for Protection and Preservation of Materials (Stone and Wood) Belonging to Cultural Heritage

Mariaenrica Frigione [1,*] and Mariateresa Lettieri [2]

1. Department of Engineering for Innovation, University of Salento, 73100 Lecce, Italy
2. Institute of Archaeological Heritage—Monuments and Sites, CNR–IBAM, Prov.le Lecce-Monteroni, 73100 Lecce, Italy; mariateresa.lettieri@cnr.it
* Correspondence: mariaenrica.frigione@unisalento.it; Tel.: +39-0832-297-215

Received: 4 July 2018; Accepted: 6 September 2018; Published: 10 September 2018

Abstract: In order to protect a material belonging to Cultural Heritage (i.e., stone, wood) from weathering, and in turn to preserve its beauty and historical value for the future generations, the contact with external harmful agents, particularly water, must be avoided, or at least limited. This task can be successfully obtained with the use of a protective organic coating. The use of nano-metric reinforcing agents in conventional polymeric coatings demonstrated to be a successful route in achieving better protective performance of the films and improved physical properties, even in extreme environments. The present paper would, therefore, review the more recent findings in this field. Generally speaking, when a hydrophobic product is applied on its surface, the stone material will absorb less water and consequently, less substances which may be harmful to it. An efficient organic coating should also supply wear and abrasion resistance, resistance to aggressive chemicals, excellent bond to the substrate; finally, it should be also able to guarantee vapor exchange between the environment and the material interior, i.e., the material should keep the same water vapor permeability as if it was un-protected. To regard to the conservation of wood artifacts, protective treatments for wood will preserve the material from environmental agents and biological attack. Hence, potential advantages of hybrid (organic–inorganic) nano-composite coatings for stone/wood have been found to be: Enhanced mechanical properties in comparison to the pure polymeric matrix, due to the reinforcing effect of the nano-filler; superior barrier properties (the presence of the nano-filler hinders the ingress of water and/or potentially harmful chemicals); optical clarity and transparency. It has been found that the efficacy of a nano-filled coating strongly depends on the effectiveness of the method used to uniformly disperse the nano-filler in the polymeric matrix. Furthermore, the presence of nano-particles should not impair the viscosity of the organic matrix, in order to employ the conventional techniques of application for coatings.

Keywords: coatings; nanosilica; nano-TiO$_2$; nano-clay; stone conservation; wood protection

1. Introduction

As recommended by conservation professionals of historic monuments and building, a protective product for stone surfaces must be able to: protect the stone from external agents and guarantee a high level of hydrophobicity, avoiding in particular the ingress of water, considering that water and water-soluble salts represent the main causes of degradation mechanisms; allow the transpiration of the stone, in order to avoid that the water already present in the substrate can cause further degradation; not alter the color and other optical characteristics—like gloss—of the substrate; be reversible, to permit an easy removing of the coating when it will result no more effective with no damage of the substrate. In addition, an ideal coating should be also stable, durable, easy to apply and non-toxic.

The protection of ancient stone monuments and buildings from the attack of environmental agents and atmospheric pollution is frequently carried out employing polymer-based surface treatments. Polymeric coatings, produced starting mainly from acrylic/methacrylic monomers, unsaturated polyesters, fluorinated polymers, epoxy resins, silanes and siloxanes, are able to form hydrophobic and completely transparent films, capable to limit the ingress of fluids into the stone. Among other advantages, their possibility to develop high mechanical resistance upon curing and hardening under a wide range of environmental conditions, low weight, also due to the small thicknesses required for an efficient protection, wear and abrasion resistance, good bond exerted to stone [1,2]. In some cases, they demonstrated also capacity of acting as consolidants. The limited durability, especially for acrylic-based coatings showing poor photo-oxidative stability when outdoor exposed, on the other hand, poses heavy concerns on their wider use when a long-term protection of archaeological and historical items and structures is of fundamental importance. The treatment efficacy (hydrophobic properties) of fluorinated polymers decreases dramatically with time [3].

Polymeric nano-composites based on nano-sized inorganic (aluminum oxide, clay, calcium carbonate, silica, and titanium dioxide) particles dispersed in the polymeric matrix are increasingly becoming a viable alternative to commercial polymeric coatings for stone due to their superior properties compared either to the pure polymer and to conventional micro-composites, especially in terms of barrier effect, higher resistance to temperatures, hash environments, fire and flame, and durability performance [4].

Durability of wood artifacts and structures, due to the timber microstructure and porosity, is strongly affected by exposure to environmental agents, mainly water/moisture presence, and microbial attacks [5]. Superficial protection and wood impregnation with chemical treatments are well known effective methods to limit the water absorption, improve the biological resistance and enhance the durability of wood. Among the most effective products for the protection of wood: aldehydes, ethyleneurea formulations, polyurethane, unsaturated polyesters, phenolic or melamine resins, acrylic and metacrylic monomers [6]. As already mentioned, acrylic based coatings display poor durability when the wood items/structures to protect are outdoor exposed. Furthermore, the most of protective coatings for wood are applied in solvent, thus during their application VOC (volatile organic compounds) can be released, representing a severe hazard for human beings and the environment.

Nano-filled coatings have been proposed for the protection of wood from biological and environmental attacks, being able to overcome some of the mentioned drawbacks. In addition, enhanced mechanical superficial characteristics (resistance against abrasion and wear), impact strength and fracture toughness, improved resistance to flame, fire and moisture can be achieved with the introduction of an inorganic nano-filler in a protective organic coating [6].

2. Nano-Clay

The more attractive properties of clay nano-particles as additive for coatings can be summarized in their: availability, low costs if compared to those displayed by other nano-particles, low toxicity and concerns for environment, good reactivity and catalytic properties [7]. Furthermore, the addition of nano-clay in polymeric coatings confer them remarkably improved mechanical properties (i.e., stiffness and strength) and hydrophobicity characteristics, as well as thermal and flame resistance [8], as a consequence of unique interfacial properties generated at nano-metric level. Due to the great specific surface area of nano-clay, furthermore, high barrier characteristics can be also achieved, important features for protective coatings for stone surfaces. The incorporation of nano-clay at low loadings, i.e., not greater than 10%, in organic (especially acrylic-based) coatings has been found to appreciably enhance the properties and the durability also of wood. Due to their small size, in fact, the nano-particles are able to easily penetrate into the wood cells.

Referring to the transparency and optical properties, the size of nano-particles, smaller than the wavelength of visible light, reduces the chance of light interacting with the nano-clays, especially if they are well dispersed in the organic matrix. This allows nano-filler to be introduced into a coating without impairing its transparency. Several studies have demonstrated, in fact, that, if limited amounts of nano-clay particles are appropriately dispersed in a polymeric matrix, good optical clarity can be achieved. However, the introduction of nano-clay at high loading (i.e., 10 wt %) in a polymeric matrix leads to a reduction of the degree of transmitted visible light through the coating with respect to the purely polymeric film, due to the light scattering by clay nano-particles, possibly aggregated [9,10].

In relation to the glass transition temperature (T_g), an increase of this property has been registered in clay nano-composites, attributed to a restriction in macromolecular mobility due to the presence of silicate nano-particles [11].

2.1. Methods of Preparation and Dispersion

The effectiveness of a protective coating depends on its chemical nature, its composition, comprising the presence of suitable additives, and on the production and application methods.

The traditional techniques employed to obtain micro-composites (i.e., blending/extrusion in melt state or in solution, in situ polymerization) can be extended also to clay nano-composites provided that an adequate dispersion of the nano-filler is achieved. In the case of in situ polymerization, the intercalation of polymer macromolecules into the clay galleries is followed by a thermo/UV-activated processes, i.e., a free radical or a ring-opening polymerizations. Photo-polymerization is defined as the cure process of a thermosetting resin induced by UV/solar light. This fairly new technology is economical and environmentally-friendly, since solvent-free mixtures can be fast hardened after their deposition as low-thickness coatings on a substrate.

Generally speaking, the production and application of a nano-filled coating presents some difficulties and requires the identification of proper procedures, with possibly greater processing temperatures and times. In the case on montmorillonite (abbreviation: MMT), the preparation method must guarantee the nano-filler exfoliation inside the matrix resin.

As an example, the most used procedure in the preparation of the nano-filled coatings is the dispersion method, where both, the nano-filler and the polymeric matrix, are solubilized in a solvent; the solvent is, eventually, removed through evaporation. In order to promote the dispersion of the nano-filler, this procedure can be enhanced through magnetic/mechanical stirring or sonication in an ultrasonic bath at appropriate temperatures. It has been demonstrated that dispersed nano-clays introduced in a mix of acrylate resins using ultrasonication techniques supply to the un-filled coatings enhanced mechanical properties [12]. Nevertheless, the production method was quite long and difficult to extend to industrial scale.

In the case of coatings based on un-saturated polyester resin and functionalized MMT powder, it was found that the performance of the MMT nano-filled coating were appreciably affected by the mixing time, being the influence of this parameter of processing comparable with the effect of the MMT content [13].

Since a good dispersion of the nano-particles in the polymeric matrix is one of the key factor for a true enhancement of final properties and performance of the matrix [14], a proper modification of the hydrophilic surface nano-particles could be necessary. In such way it is possible to increase the compatibility between a nano-sized clay and the polymeric matrix and to obtain an homogeneous and more stable dispersed nano-filled suspension in liquid state. The superficial modification of the nano-clay, as well as its loading into the polymeric matrix, can also have a remarkable effect on the extent of cure of the polymeric coating.

In some case, water-based dispersions of nano-particles are preferred in order to avoid the aggregation of nano-particles both in the wet state of the coating and in the hardened state of the protective film.

2.2. Relevance of Viscosity for the Application

The assessment of the viscosity of the un-cured coating is essential to identify the proper procedure (for instance, using a brush, by spray or capillary rising) to apply the liquid product on the substrate to protect. In addition, the viscosity can also be affected by the preparation procedure employed to produce the nano-filled coating [13]. On the other hand, by increasing the resin viscosity, it becomes more difficult to produce thin coatings. Therefore, rheological studies must be performed in order to quantify the effect of the presence of the nano-filler on the viscosity of the liquid coating and to optimize the parameters for the production technique.

2.3. Montmorillonite

Among smectite class of aluminum silicate clays (i.e., those possessing layered silicate particles), montomorillonite (MMT), whose structure is shown in Figure 1, has gained the major attention both within the academia and industrial research for the production of polymer-based nano-composites, giving rise to well-developed intercalated nano-sized structures [15,16].

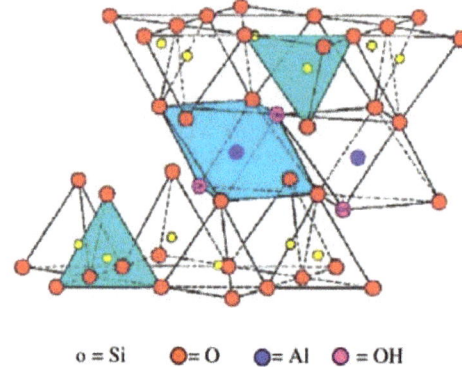

Figure 1. Molecular structure of montmorillonite. Reprinted with permission from [16]. Copyright 2008 Elsevier.

A hybrid composite, composed by an organic matrix and MMT nano-filler, can display the three different following morphologies [17]: (i) incompatible one, when clay tactoids are not dispersed in the organic matrix; (ii) intercalated morphology, in this case the layered silicates are expanded in such a way that the polymer macromolecules can enter them, still surviving an ordered structure; and (iii) exfoliated morphology, when the single layered silicates are completely separated and uniformly distributed in the organic matrix, depending their average distances on the clay loading (see schemes in Figure 2). This latter morphology has been found to supply better final properties to the polymer/nano-clay systems [18].

Layered silicate nano-particles display intrinsically improved properties, such as: (i) a high aspect ratio, responsible for their platelet aspect; (ii) a large specific surface area (750–800 m^2/g); and (iii) greater Young modulus with respect to nano-sized silica particles [19]. The structure of layered silicate nano-particles when exfoliated into the polymer, i.e., high aspect ratio and high surface area, are responsible for the enhanced properties of their nano-composites, in particular barrier capability. The silicate type (structural features) and its loading in polymeric matrix will affect the final properties and performance of the resulting nano-composite. A small quantity (typically not greater than 5 wt %) of layered nano-clays can guarantee a large interfacial area with the polymeric matrix, provided that the nano-filler is properly dispersed into the matrix [15]. An appropriate chemical modification of the hydrophilic surface of MMT nano-clays is frequently carried out with the aims to favor their compatibility with different polymers with the additional advantage to increase the interlayer distance, thus facilitating the ingress of the macromolecules during the dispersion process [15].

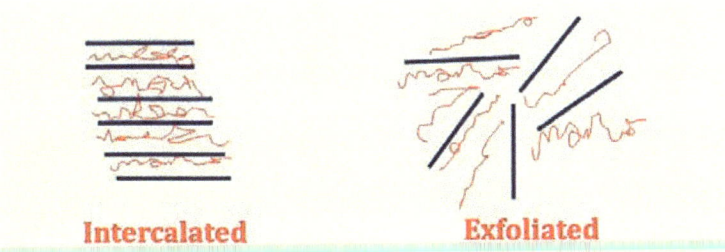

Figure 2. Schematically illustration of intercalated and exfoliated morphologies displayed by polymer/layered silicate nano-composite.

2.3.1. Montmorillonite/Stone

Only few examples on the use of polymeric coatings based on montmorillonite for the protection of stone surfaces are reported in literature.

Experimental coatings produced starting from a commercial formulation, based on a blend of acrylic and vinylidenfluoride-based polymers, and different amounts (not greater than 4 wt %) of an organically modified MMT (i.e., Cloisite 30B) were tested as protective/consolidating agents for a porous stone typical of the Campania region in Italy, i.e., the Neapolitan yellow tuff [20]. It was found that the addition of low amounts of the organoclay nano-filler into the commercial polymeric formulation allows to achieve reductions in both water absorption and water vapor permeability. Enhanced mechanical properties and abrasion resistance were also observed, suggesting that the nano-composites display also consolidating capability. In addition, the chromatic appearance of the stone was not altered by the MMT-based coating.

Water-based nano-composite, obtained by dispersing low amounts (up to 5 wt %) of MMT (neat or organically modified) nano-particles into a cross-linked fluorinated polyurethane based on perfluoropolyethers, were analyzed as possible anti-graffiti coating for Leccese stone, a very porous stone characteristic of Apulian region [21]. It was found that organically modified MMT when introduced (in particular at 3 wt %) in the polymeric matrix is able to supply a high hydrophobicity to the stone and a stable anti-graffiti effect for black acrylic spray paint; this latter, in fact, can be easily removed from the stone using MEK (methyl ethyl ketone) solvent, leaving the stone surface substantially unaffected from an aesthetical and morphological point of view, even after repeated staining/cleaning cycles (as shown in Figure 3). Unsatisfactory results were found when the leccese stone surface was staining by a black ink permanent marker. Irrespective to the MMT nano-filler employed, in fact, the coatings were able to protect the stone only for the first staining/cleaning cycle. After repeated cycles, on the other hand, the marker is able to penetrate the nano-filled coatings and the solvent employed to clean the surface, i.e., IPA (Isopropyl alcohol), is no more able to remove the black ink. Furthermore, a noteworthy drawback observed for the nano-composite coatings was a significant reduction in water vapor permeability brought about by their application on Leccese stone. However, this inconvenience was mainly attributed to the cross-linked nature of the fluorinated polyurethane able to hinder to a certain extent the transport of water vapor exiting from the stone.

In a different study, montmorillonite clay (up to 7 wt %) was added to a biodegradable polymer (poly-L-lactide, PLA), obtaining bio-nano composite coatings proposed for the protection of marble against pollution [22]. With the application of the protective coating, significant improvements in hydrophobicity were registered on the marble, both capillary water absorption and transport of water vapor were reduced, both attributed to the presence of exfoliated MMT nano-clays hindering the water movements inside the polymeric coating. The chromatic appearance of the surface remaining unchanged with the application of the bio-nano composite coating. In addition, a greater resistance of the marble to acidic atmosphere was registered. Finally, due to the nature of PLA matrix, it is expected a limited biodegradability of the MMT coating, that will suggest a reversibility of the treatment.

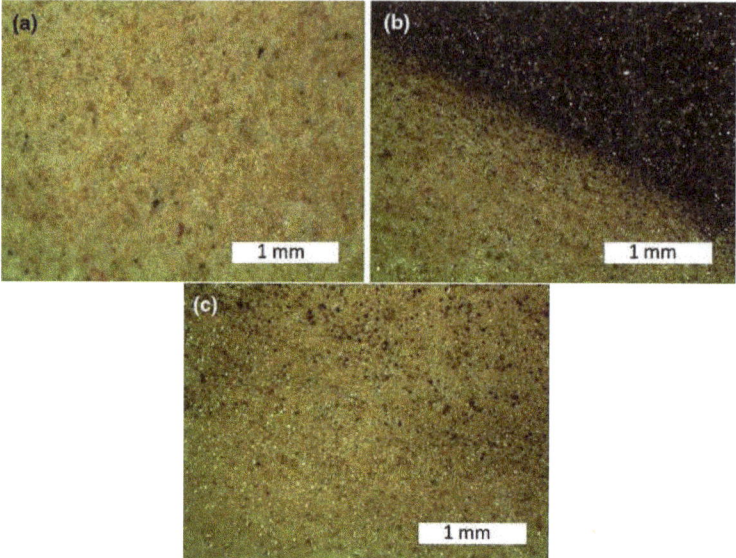

Figure 3. Micro-graphs at reflected light showing Lecce Stone surface: (**a**) after treatment with nano-composite containing 3 wt % of organically modified montmorillonite; (**b**) after staining by spray paint (upper right area in the picture); and (**c**) after 8 consecutive staining/cleaning cycles. Reprinted with permission from [21]. Copyright 2014 Springer.

2.3.2. Montmorillonite/Wood

Several are the recent examples of MMT-based nano-composite coatings for the protection of wood surface reported in literature. Even though most of the experimented hybrid systems are proposed for wood employed in modern floor, furniture or topping, such products could be extended also for the protection of wood artifacts and timber of valuable ancient structures.

The effect of dispersion method employed to introduce modified nano-composites based on MMT (i.e., Cloisite modified with a quaternary ammonium salt) in acrylate oligomers to produce UV-cured coatings for sugar maple wood [9]. The effect of process type, in particular of the quality of the clay dispersion, and of clay loading (1–10 wt % of modified Cloisite) on surface-mechanical properties and optical characteristics of the coatings was assessed. Both these parameters were found to appreciably affect the mechanical and optical properties of the nano-coatings. It was found that at small clay loadings (not greater than 3 wt %), with a suitable dispersion technique, nano-composite UV-cured acrylic coatings with significant mechanical and optical properties can be produced for the protection of wood, mainly for flooring applications.

In other studies, three different commercial organically modified Montmorillonite (i.e., Cloisite 10A, Cloisite 15A and Cloisite 30B) were added (1 and 3 wt %) to an UV-cured epoxy-acrylate oligomer intended as protective coatings for wood furniture [19,23]. The Authors assessed that not all the clay nano-fillers were well dispersed in the acrylic matrix, and only C10A and C15A give rise to an intercalated morphology in the UV-cured nano-composites, irrespective to the nano-filler content. The kind and amount of added organo-clays affected both the water vapor transmission rate and the optical transparency, both generally decreased upon addition of MMT nano-particles. C10A seemed to supply the best performance to the organic coating. The presence of low amounts of organo-clays is also able to enhance the mechanical properties and glass transition temperature of the pure resin, particularly for nano-composites based on C10A. It can be concluded that the presence of low amounts of a suitable organo-clay in an acrylic matrix can confer better mechanical properties and an increase of

T_g to the UV-cured film. The resulting organic nano-filled coating also displays a limited deformability, which represents an advantage in comparison to the inorganic protective products.

The effect of nano-clays addition to nano-composites based on TiO_2/acrylic transparent coatings applied on Norway spruce was assessed in several studies [7,24,25]. Accelerated weathering tests assessed that the presence of low amount (1 wt %) un-modified nano-clay (bentonite, i.e., colloidal clay consisting primarily of MMT) in combination with TiO_2 nano-particles is able to supply enhanced resistance towards UV radiation to the acrylic coatings, delaying the photo-degradation process of protective coatings, the latter keeping transparency upon outdoor exposure. At slightly greater amounts (i.e., 3 wt %), the sole presence of nano-clay in the water based acrylic coating is still able to retard to a limited extent the degradation of wood surface.

3. Boehmite

Boehmite (c-Al_2O_3) nano-particles are colloidal plate-like crystals. As shown in Figure 4, Boehmite crystals, characterized by a high anisotropy, consist of double layers of oxygen octahedrons with a central Al atom [26]. An important feature of Boehmite is that it may have a very high specific surface area (>300 m^2/g), suggesting its use as nano-filler in polymeric materials. Furthermore, Boehmite nano-particles are largely available on market at reasonable costs; their surface can be easily treated (with proper coupling agents) to assume hydrophobic/hydrophilic character, enhancing their dispersion in various polymeric matrices.

Upon their addition to a polymeric resin, organically modified Boehmite nano-particles were found to increase the T_g, the hardness and mechanical strength of a cured epoxy resin [27], being the latter all attractive characteristics for a protective coating for stone or wood.

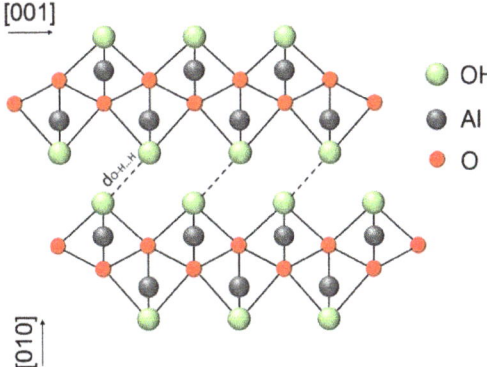

Figure 4. Molecular structure of Boehmite. Reprinted with permission from [26]. Copyright 2017 John Wiley & Sons.

Referring to the production techniques, the addition of organo-modified Boehmite (amounts in the range 5–10 wt %) to a UV-cured cycloaliphatic epoxy resins, to obtain in situ photo-polymerized nano-composite coatings, resulted in a reduction of the epoxy resin reactivity. This un-desirable effect was attributed to the light absorbance of the nano-particles due to scattering [28]. However, the final properties of the coating, i.e., high transparency and glass transition temperature, were suitable for protection applications in Cultural Heritage field, especially when modified Boehmite was added to the epoxy matrix at low loadings. On the other hand, when organically modified Boehmite (up to 5 wt %) was added to a DGEBA (diglycidyl ether of bisphenol A) epoxy, both the cure kinetic of the resin and its degree of cross-linking were increased [27]. Similar results were found for the same nano-filler, which was added to a siloxane-modified methacrylic resin: the conversion of reactive species in the organic matrix was found, in fact, to be increased by the presence of Boehmite nano-particles, remaining fairly unaffected the kinetics of cross-linking [14].

Referring to processability characteristics, a general increase in the viscosity of the polymeric resin was observed with increasing the amount of organically modified Boehmite nano-particles, especially at very high loading amounts (up to 20 wt %), irrespective to the chemical nature of the epoxy matrix [14,27,29].

3.1. Boehmite/Stone

UV/sunlight curable nano-composites suitable for stone protection were produced starting from an experimental siloxane-modified methacrylic formulations with the addition of organo-modified boehmite (OMB) [30]. The presence of nano-particles uniformly dispersed in the thermosetting polymer allows to increase its hydrophobicity, hardness and superficial resistance to scratch. The nano-composite coating displays complete transparency and slightly greater T_g as compared with the unfilled UV-cured modified methacrylic resin. An increase in thermal conductivity was also registered upon addition of OMB nano-particles into the polymeric resin [31].

Referring to durability characteristics, after two subsequent cycles of accelerated aging (duration 100 h each), the surface functional properties of the protective coatings (T_g, contact angle, hardness) slightly increased and, then, remained unaffected. The presence of nano-particles was able to reduce, but not to completely avoid, the yellowing effect due to prolonged exposure (four cycles) to the UV-lamp in the accelerated weathering machine.

The experimental sunlight-cured nano-composite was, then, applied to two different calcareous stones (i.e., Leccese and Gentile stones, typical of Apulia region) in order to assess its performance as protective coating for stone substrates [4,31]. The organo-modified Boehmite-based nano-composite coating showed a noticeable hydrophobic character (as shown in Figure 5), demonstrated to be able to reduce appreciably the ingress of liquid water in both experimented stones. It was completely transparent and no chromatic variation was registered on both stones upon its application. The good breathability assessed for the experimental coating applied on both stones is maintained even after prolonged outdoor exposure or an accelerated aging procedure [32]. The good protective properties and durability performance assessed for the nano-composite coating were found even greater than those measured on two commercial organic water-borne hydrophobic products. A further advantage of the OMB-based coating over commercial products is the complete absence of any solvent in the liquid uncured formulation, so it can be truly considered as a "green" product.

Figure 5. Measurements of contact angle performed on (**a**) Leccese stone and (**b**) Gentile stone specimens, both treated with Boehmite-based nano-composite coating. Reprinted with permission from [32]. Copyright 2017 Elsevier.

3.2. Boehmite/Wood

The nano-composite UV-cured coating, produced starting from a siloxane-modified acrylic formulation with the addition of an organo-modified Boehmite, successfully experimented on stone was, then, proposed for the protection of walnut wood [14,33]. A high penetration depth, comparable to that measured on commercial products (based on linseed oil) was assessed for the nano-filled modified-acrylic liquid formulation. The wood surfaces treated with the nano-composite resulted highly hydrophobic and absorbed smaller amount of water with respect to commercial coating. The transparency of the coating film remained unaffected by the presence of OMB, with a small chromatic alteration of the wood substrate, even lower than that measured for the commercial product. Along with an enhancement in surface hardness, the presence of a low amount (at 3 wt %) of organically-modified nano-particles produced an increase in T_g in comparison to the unfilled acrylic mixture. As previously underlined, the liquid formulation is completely solvent-free, this latter representing a double advantage: the absence of any toxic-hazardous solvent and the drastic reduction of the hardening time of the coating, passing from 1 week, required to achieve the complete evaporation of solvents in the commercial product, to few (six) hours necessary to complete the photo polymerization process in the nano-composite coating.

4. Nanosilica

Several methods have been used to design and produce superhydrophobic/oleophobic stone surfaces, hierarchically structured (Figure 6) [2] and often bioinspired. The most common procedures include sol-gel processes and controlled nano-particle embedding into polymer matrices. This latter is usually a simple method, having low costs, because common polymers, nano-particles and solvents need to be used. The obtained hybrid polymer films, applied on the stone surface, are able to reduce the surface tension of the substrate, thus increasing the surface hydrophobic properties. Hydrophobicity can be further enhanced by increasing the surface roughness. To this aim, the addition of nano-particles has been found very useful and effective. Among them, nano-scale SiO_2 particles have been widely applied.

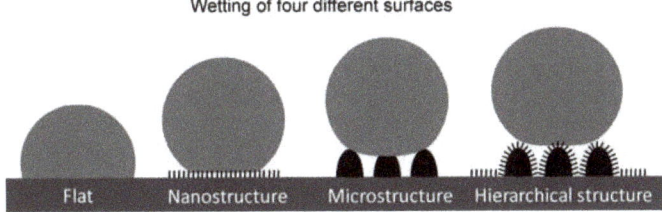

Figure 6. Schematic of four different surface structure types and their wetting behaviors. Reprinted with permission from [34]. Copyright 2016 Elsevier.

4.1. Preparation and Properties of Nanosilica Hybrids

Nano-scale silica (SiO_2) particles do not have the narrow gallery structure of layered clay, but possess a large interfacial area as long as the diameter of the particles is in the range of nano-meters, and they can be well dispersed in polymers [35]. On the other hand, nanosilica alone works quite well only as a consolidant, but exhibits poor results on carbonate substrates, probably due to the incompatibility of such product with the stone [36].

Silicon dioxide nano-particles (from 0.1% to 8%) have been added to polymer based on silanes/siloxanes [37–39], fluoropolymers [40,41], polyurethanes [42,43], polyurethane-acrylate [44], methacrylic resins [45,46], unsaturated polyester resin [47], DGEBA-amine epoxy resins [48,49]. In these hybrids, silica nano-particles usually have dimensions around 7–14 nm in mean diameter [37,38].

After the nano-particles' addition, the mixture with polymer(s) can be subjected to sonication [42,48,50,51] or vigorous stirring in order to prepare an optimal and homogeneous dispersion [37,47,52]. However, sometime the stirring procedure did not ensure nano-particles homogeneously dispersed in the polymer film; SEM (scanning electron microscopy) observations highlighted aggregates with average diameters dependent on the nano-particles concentration [38,52]. Sonication has been found useful to change the basic rheological properties of polymers by partial disintegration of the primary polymer chain structure and cavitation exerts, thus impacting on the ability of polymers to well penetrate and fill the pores of the stone substrate [51].

The preparation through a sol-gel process is usually carried out via co-condensation of compounds containing –Si(OR)$_3$ moieties (such as tetra-ethyl-orthosilicate (TEOS) [43,45]) and a polymer (polydimethylsiloxane (PDMS) [53–55], trimethoxysilyl-propyl methacrylate (TMSPMA) [56]), in the presence of a surfactant (n-octylamine [53,54,57], dodecylamine [56]). In the processes where techniques based on two-stages [56], that is sol-gel and free-radical polymerization, have been applied to obtain hybrid nano-composites, the 2,2-azobis(2-methylpropionitrile) (AIBN) and/or acids can be used as additional reagents. A preliminary functionalization of the silica nano-particles (for example with 1,1,1-Trimethyl-N-trimethylsilyl-silanamine) is sometime performed [38].

In the sol-gel process, hydrolysis reaction rate plays a critical role: when hydrolysis is sufficiently rapid to provide hydrolyzed products that can condense with the polymer component (e.g., silanol-terminated PDMS), a homogeneous organic-inorganic hybrid gel is obtained. The surfactant is able to modify the hydrolysis/condensation process, because of a change in pH of the solution which modify the rate of hydrolysis; an increased pH of the solution increases the level of hydrolysis [54].

The physical properties of nanosilica-based hybrids mainly depend on the synthesis procedure, which affects the specific surface, the dimensions of the primary particles and aggregates, and the coating' porosity. The differences between these physical characteristics are strictly correlated to the organization of the silica particles (as aggregates or agglomerates) [56].

Nanosilica can be observed in aggregated state with the different size and shape, depending on the type of matrix. In fact, the dispersion of the nano-particles depends on the extent of interactions between them as well as on the polymer/nanoparticles interactions [58]. Increasing the nanoparticles' loading results in an increase in both the size and number of the nanosilica aggregates [59] and a decrease in the inter-aggregate distance, up to obtain a fractal structure [58,60]. In these cases, nanoparticles in agglomerates interact directly with each other with this attraction prevailing over all the interfacial interactions in the system [60].

The materials synthesized using a surfactant exhibit high surface area, high pore volume, and a structure having uniform mesopores [53,54]. Since the capillary pressure in the gel network during drying is inversely proportional to the gel pore size, the coarsening of the gel pores promotes a lower capillary pressure, thus, preventing cracking. The polymer itself can prevents cracking, by increasing the flexibility of the silica skeleton and the shrinkage during drying [54].

The addition of silica nano-particles causes some changes in the rheological properties of the neat polymer. Nanosilica increases the viscosity values, imparts pseudoplasticity (i.e., the viscosity decreases by increasing the shear rate), and gives thixotropy (i.e., the viscosity increases by increasing the time after shearing) [47]. The increase in viscosity is mainly related to the aggregation of the primary particles [47]. For this reason, the viscosity increases with the nanosilica content, in particular for amounts of nanosilica higher than 2 wt %. Also the choice of the polymer can influence the viscosity. For example, the addition of PDMS to the starting sol can promote a slight increase in viscosity [54].

The incorporation of nanosilica increases the T_g because the nano-sized particles act as crosslinkers as a consequence of the nano-filler-polymer interactions [47,56,61].

4.2. Nanosilica Hybrids on Stone

Addition of silica nano-particles at low concentrations (1–2% w/w) results in the formation of film on stone with a continuous structure having nano/micro-scale roughness (Figure 7), sometimes exhibiting a two-length-scale hierarchical structure on the surface which is the reason for the superhydrophobic nature of these coatings [39].

The higher the nanosilica consistency the more the possibilities of agglomeration phenomena, which took place for nanosilica content above 2% [50]; for similar or higher amount of nano-particles, cracks and grooves are observed in some places of the layer [37,50,57] (Figure 8).

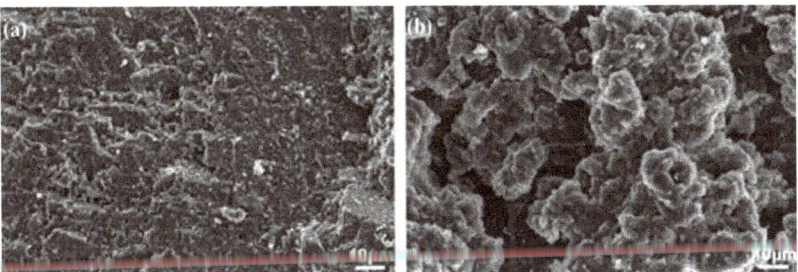

Figure 7. SEM images of (**a**) untreated marble and (**b**) marble covered by a superhydrophobic, water repellent composite (the silica/siloxane mass ratio was 0.4). Reprinted with permission from [39]. Copyright 2013 John Wiley & Sons.

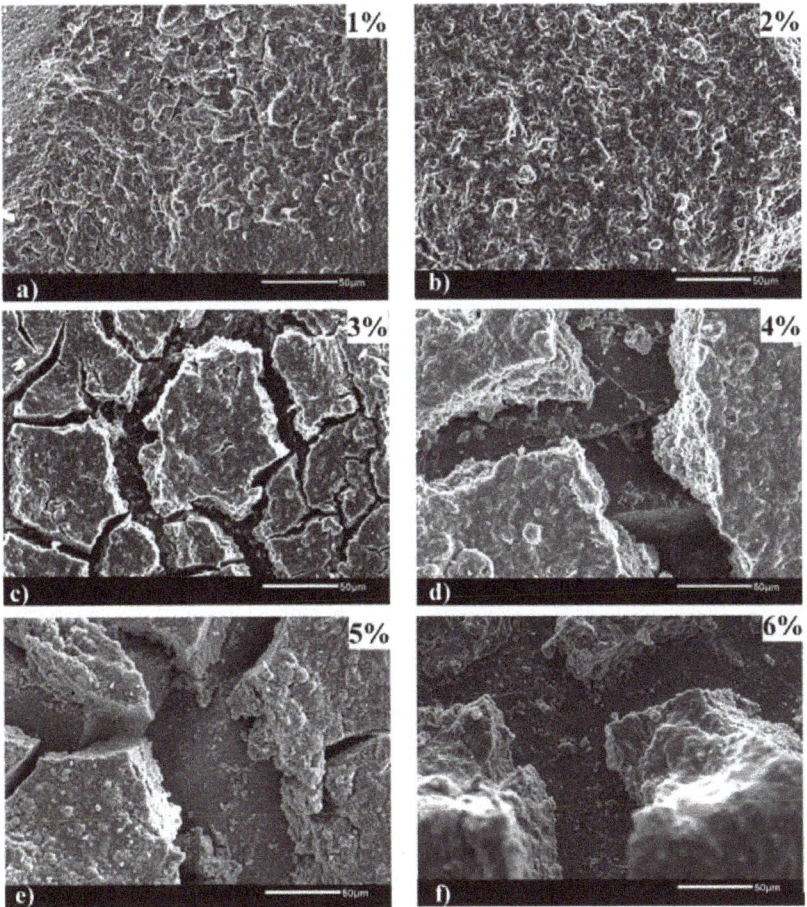

Figure 8. SEM images of marble surfaces coated with siloxane with SiO_2 nano-particles at increasing loading (from (**a**) to (**f**)). In the upper right corner of each image the nano-particles' concentrations (wt %) is reported. Reprinted with permission from [37]. Copyright 2016 Elsevier.

The stone treated with a hybrid obtained by sol-gel process usually exhibits crack-free mesoporous coating on the surface; furthermore, numerous bridges linking together the grains of minerals in the stone can be observed [41,53,54] (Figure 9).

A cracked morphology affects the wettability of the surface. In particular, a strong decrease in contact angle is measured using low surface tension liquids (e.g., oil): drops of these substances sink into the grooves and gradual transition from the non-sticking Cassie-Baxter state to the Wenzel state occurs [37,40]. In addition, the polymer film in-between aggregates is thinner or even disrupted, leading to a decrease in the efficiency of stone protection from water capillary absorption [52].

A significant increase in contact angle is measured and super-hydrophobic behavior is observed with the addition of 1.5 wt % [50] of nano-particles; sometimes, lower or higher amounts than 1.5% decreased the contact angle values [50]. It is to take into account that nanosilica alone is typically hydrophilic [62]. The nano-dimensions of silica particles are essential for the super-hydrophobization of the stone surface. In fact, in experimental studies where silica micro-particles (mean diameter 3 μm) in polymer were applied, the super-hydrophobic effect was not achieved and substantially lower static contact angles (120° instead of 161°) were measured [52].

Figure 9. SEM image of a biocalcareous stone treated with an organic-inorganic hybrid (obtained by co-condensation of TEOS and polydimethylsiloxane): (**a**) coating on the stone grains; (**b**) nano-material bridges linking together the grains of the stone. Adapted with permission from [54]. Copyright 2010 American Chemical Society.

Contact angle values obtained on surface treated with nanosilica hybrids obtained via sol-gel processes showed a significant enhancement in hydrophobicity, up to super-hydrophobicity. Because the water contact angle cannot be increased beyond 120° by a purely chemical process on a smooth surface, the high contact angles (>140°) are due to the combined effect of a low surface energy of the polymer and surface roughness [54]. However, in these hybrid materials super-hydrophobicity is not always achieved [53]; this worse behavior is observed where the treatment does not affect surface roughness [43].

Highly water-repellent properties are observed along with of super-hydrophobicity on stone surfaces treated with nanosilica hybrid polymers [39,52,63]. This property has been confirmed by the evaluation of the water contact angle hysteresis, which is the most adequate parameter for the assessment of water-repellency. A highly water-repellent surface exhibits a low water contact angle hysteresis (<10°), that is a low difference between the advancing and receding contact angles. In these conditions, a water droplet can move with little applied force and easily rolls off from the surface (Figure 10).

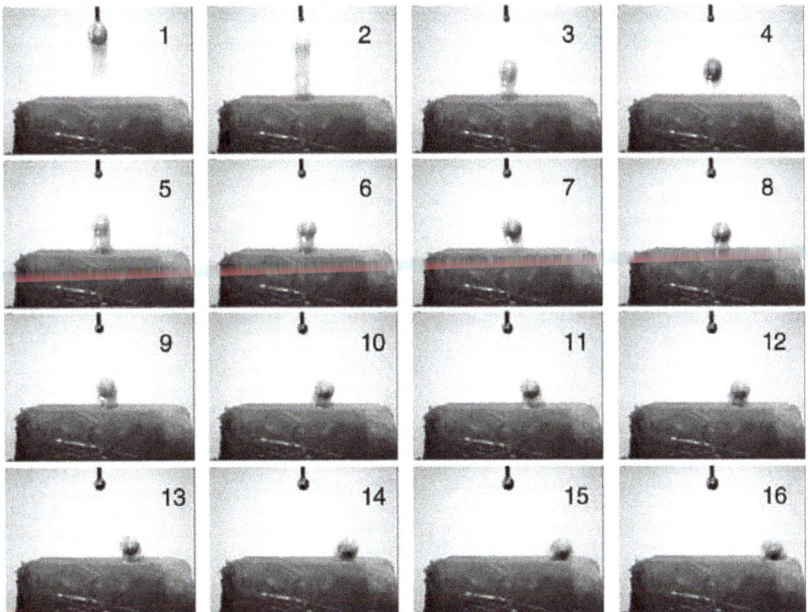

Figure 10. Consecutive images of a water droplet bouncing and rolling off a Pentelic marble treated with siloxane and 1.5% w/v silica nano-particles. Reprinted with permission from [52]. Copyright 2009 Elsevier.

Although high hydrophobicity (up to water repellence) is observed, water absorption by capillarity is sometime not adequately reduced, partially in contrast with the static contact angle results [38].

The addition of silica nano-particles to the polymer does not significantly reduce the water vapor permeability [51], in particular for porous stone materials [40,64]. Reduction in breathability is measured at high nanosilica concentration, being this reduction associated with the density of the aggregates formed on the surface of the treated stone [52]. Usually, the hybrids obtained via sol-gel do not completely block the porosity, as a consequence the breathability of the stone after the treatments is only slightly reduced [53].

The mechanical characterization has demonstrated a positive influence of the addition of silica nano-particles on impact resistance [47], scratch resistance [61], as well as on surface hardness, because of capacity of coating reorganization and reconstruction during the mechanical tests [51]. Where the products penetrate the stone less deeply, the surface resistance, rather than the compression strength, has been found more appropriate for evaluating the effectiveness of the consolidation for this stone [53].

Nanosilica-based polymer hybrids show a good durability over time, either in accelerated [55,56] or natural aging. In an experimental study, after 5 months of outdoor exposure, water contact angles were reduced, but the treated stone surface remained super-hydrophobic [52]. Amounts of silica nano-particles higher than 2% allow a suitable resistance in terms of weight loss to freeze-thaw cycles and saline (NaCl) action.

With regard to the durability of these hybrid coatings, release of nano-material by matrix degradation in outdoor uses is a topic of great concern because the polymer binder in a nano-coating can be prone to degradation under the UV radiation of sunlight. Due to the matrix degradation, the nano-particles could be exposed on the coating surface and become more likely susceptible to be released into the environment [42], for example by rain action [49]. The rate of chemical degradation of the hybrid polymer can be lower than for the neat polymer, indicating that the nanosilica particles are able to photostabilize the matrix. In the degradation process, silica nano-particles have been observed to accumulate and cluster on the nano-coating surface with increasing UV exposure time (Figure 11); the release take place not randomly but where a critical concentration is reached [42,48].

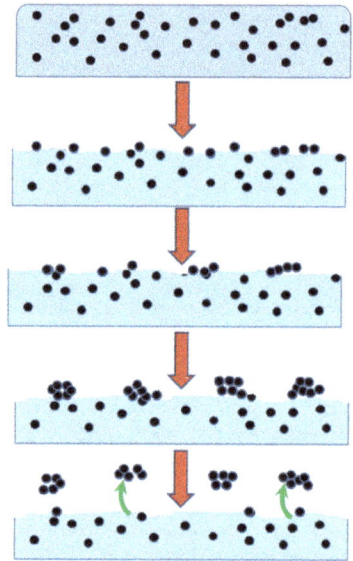

Figure 11. Model for release of nano-SiO$_2$ fillers from a hybrid composite: matrix degradation, formations of particle clusters, and release of particle clusters under increasing UV exposure (from up to bottom). Reprinted with permission from [42]. Copyright 2016 Springer.

4.3. Nanosilica Hybrids on Wood

Silica nano-particles have been proposed also as nano-reinforcement in surface coating applications to protect surface of wood substrates.

Addition in UV-photocurable polyurethane acrylate dispersions [65], epoxy acrylates [66,67], acrylic resins [68] has been studied. Photoinitiators, defoamers, surfactants and/or rheological modifiers complete the formulation. Silica nano-particles, added in amounts ranging from 0.5% to 5%, usually have a particle size of 10–50 nm. Their surface modification with methacrylsilane groups is sometimes suggested [65–67].

Despite the presence of nano-particle aggregates, an improvement of scratch resistance of the applied coatings is measured [69]; an increase of more than 200% of scratch resistance was reached using only 1 wt % of nanosilica [65]. The improved scratch resistance is attributed to the reduction of free volume of the hybrid [65]. In fact, as already reported in the previous sections, the addition of nano-particles tends to hinder macromolecular chains mobility at the interface around them. Therefore, in the coating, free volume decreases and the T_g increases.

Reduction of gloss, attributed to an enhancement of surface roughness, and good adhesion to the substrate are also obtained [65]. Adhesion is primarily due to the penetration by capillarity of the liquid coating into the pores of wood and to the subsequent solidification of the polymer. Therefore, a suitable adhesion depends on both porosity properties of the wood substrate and rheological properties of the liquid coating. High amounts of nano-particles increase the viscosity of the applied products because of aggregation effects [66]; as a consequence the adhesion strength is reduced [65]. However, the presence of nano-particle aggregates does not strongly affect the adhesion properties probably because their dimensions are small enough to make it possible the penetration into the pores of wood [65].

Super-hydrophobicity is achieved also on wood substrates where a hierarchical structure of the coating with roughness at two length scales is observed. In some cases, super-hydrophobicity is retained also after washing and abrasion test [70].

5. Nano-TiO$_2$

TiO$_2$ nano-particles have been extensively applied to obtain multifunctional photocatalytic coatings on architectural stone surfaces [71].

It is well know that materials based on TiO$_2$ nano-particles, thanks to their photocatalytic properties, are able to keep surfaces self-cleaned and to prevent the formation of biofouling. However, for the maintenance of materials in Cultural Heritage, the application of nano-TiO$_2$ dispersion alone seems to be a strategy not effective enough, especially because porous and rough substrates are treated [72]. The nano-particles poorly adhere on the stone surface, tend to penetrate in the porous microstructure, and are easily removed by the mechanical action of wind and rainfall, thus leading to a significant decrease of the effectiveness of the treatment [73]. The application of higher amounts of TiO$_2$ cannot yield better results because the risk of chromatic variations and micro-cracks in the coating is increased. Therefore, the recent approaches are focused to insert these nano-particles inside a polymeric matrix, in order to increase the surface roughness without changing the main characteristics of the material, such as permeability and transparency, and the substrate morphology [72,74]. By adopting an appropriate procedure, it is possible to harmonize the characteristics of the TiO$_2$ and the properties of polymer materials, obtaining nano-structured inorganic-organic hybrid coatings that protect against water, as evidenced by high contact angles, and photodegrade pollutants when illuminated by sunlight. Among the three crystalline forms of titanium dioxide: rutile, brookite and anatase, the last one is generally considered the most active photocatalyst.

5.1. Preparation and Properties of Nano-TiO$_2$ Hybrids

Titanium dioxide nano-particles have been added to polymer based on acrylic [75–77], siloxanes [74,77], fluorinated polymers [74,75,78,79], polyurethanes [80]. Organic solvent [73,75] or aqueous suspension [73–76] can be employed. The used TiO$_2$ nano-particles, usually having particle mean diameter of about 20–50 nm, are commercially available products or are synthesized through sol-gel method by hydrolysis of titanium isopropoxide in acid condition [14]. In the latter case, a suitable amount of a surfactant is added to the sol in order to disperse the nano-particles and avoid aggregation processes. In fact, this method allows the preparation of a product with a small percentage of brookite and the anatase phase predominant (evaluated by XRD and Raman); in addition, the crystallites size of anatase particles, estimated by TEM, can be in the 4–6 nm range [81].

The process of mixing of polymer and TiO$_2$ can be obtained by stirring or sonication [82]. However, the low chemical affinity between organic polymers and inorganic particles makes the mechanical methods of dispersion insufficient, while a chemical approach is usually more effective [83]. The addition of coupling agents (tetraisopropoxide or TEOS) to the film formulation [83] or nano-particles' functionalization [81] are the preferred methods.

The addition of TiO$_2$ nano-particles can induce increases in T_g mainly because the inorganic nano-phase reduces the mobility of the polymer chain segments [80]. This aspect can be particularly important for applications in the field of Cultural Heritage, where coatings with close or lower the room temperature are too much soft for working and moreover are inclined to pick up dust and dirty from the environment.

5.2. Nano-TiO$_2$ Hybrids on Stone

EDS investigations well follow the penetration depth of treatments on stone, due to the fact that titanium is not a component of most stone materials (in particular, marbles and calcarenites) [75]. Although it can be assumed that the polymer spreads at least as much as TiO$_2$, sometimes a lower penetration of the smaller nano-particles has been found. This leads to a higher nano-particles/binder ratio on the surface, which improves the efficacy of TiO_2 [75].

It is worth noting that the penetration depth of TiO_2 does not necessarily matches that of the polymer, since the size of these two components are different. The result mainly depends on the porosity of the substrate; the greater the porosity, the higher the penetration depth. In highly porous stone, TiO_2 hybrid polymers have been found up to 3 mm under the surface [75,76].

A deeper penetration is useless for the coating performance in terms of hydrophobicity, but can slightly inhibit the photocatalytic properties of the TiO_2 nano-particles on the surface. Conversely, a certain level of penetration can assure an adequate anchoring of the coating to the stone, which assures implementation of hydrophobicity and durability [75]. Additionally, greater roughness and porosity promote adhesion of the polymer to the substrate [75].

The presence of low amounts of TiO_2 in nano-composites are difficult to detect by FTIR [73]. The appearance of an absorption peak at about 700 cm^{-1} is related to Ti–O stretching; in addition, the increase of the OH stretching band (in the range 3400 and 3000 cm^{-1}) and OH bending signal (around 1630 cm^{-1}) is due to emisorbed hydroxyls groups bonded to TiO_2 [74]. In silicon-based hybrids also the slight broadening of the peak at about 950 cm^{-1} is observed, due to the formation of Si–O–Ti interactions between the polymer matrix and the nano-particles [74].

As SEM and TEM observations highlighted, no aggregation of nano-particles occurs in polysiloxanes-based treatments [73], even for high TiO_2 concentrations (44 wt %) [74]. TiO_2 nano-particles tend to aggregate in micrometric clusters in fluoropolymer-based nano-composite, instead, due to the low affinity between nano-particles and the water dispersion of fluoropolymers [74].

Where the coating induces or enhances the roughness, often due to the nano-particles' disposition, a different aesthetic appearance of the surface can be observed [75]. In addition, it must be taken into account that TiO_2 nano-particles are usually white in color, so that they can produce a lower color variation on whitish stone, such as marbles [75]. Furthermore, this whitening effect can have a beneficial influence, balancing the yellow color of some polymer [73,74]. The capability of producing transparent polymer coatings modified with TiO_2 nano-particles depends on the particle dimensions and on their effective dispersion in the polymer medium. Particle size lower than 50 nm allows that light is transmitted and the final coating is transparent [83].

When TiO_2 nano-particles and the hydrophobic polymer are mixed together, the final water contact angles remarkably increase as the TiO_2 amount increase in the hybrid formulation [74,78]. Also water repellency can be achieved, with water drops bouncing and rolling off the treated surface [84].

An increase of nano-roughness of the treated surfaces, with the reduction of the surface free energy, increases by increasing the TiO_2 concentration and higher values are related to the reduction in wettability of the treated surface [74]. The changes in nano-roughness are more relevant on stone materials characterized by low roughness values, such as marbles, while this parameter is slightly affected where the original roughness is already high [73]. In this latter case, the distribution of the hydrophobic coating than the presence of nano-particles affects the wettability of the substrate.

A high hydrophobicity of treated stone surfaces has been measured by using rod-shaped TiO_2 nano-crystals in oleic acid [85]. In this study, surprisingly, in the absence of TiO_2 a hydrophilic behavior was observed. This result has been explained considering that molecules of oleic acid, when not coordinated to the TiO_2, can randomly orient on the stone sample. The TiO_2 nano-rods are able to coordinate the molecules of oleic acid that expose their hydrophobic moieties toward the solid–air interface, thus decreasing the wettability of the stone surface.

Capillary water absorption is usually positively influenced in presence of TiO_2 nano-particles [73,74,76]. However, this parameter can be found in disagreement with contact angle measurements performed on the same specimens [75,79], even if low amounts of absorbed water are expected to be associated with high contact angle values. However, the results provided by the contact angle are related to an "instantaneous" hydrophobicity and take into consideration only the interface drop-surface. Conversely, the absorption test verifies the long-term water resistance measuring the quantity of water absorbed over the entire area of the sample [75].

It is worth noting that those treatments that show a better penetration exhibit a better hydrophobicity at long-term [75], but a worse behavior in terms of photocatalytic activity (biocidal effect and stain discoloration) [76].

Different behaviors in terms of water vapor permeability have been observed in the presence of TiO_2 nano-particles within the polymer. In some cases, the use of a hybrid have significantly reduced the vapor permeability (up to 60% [73]), indicating that the product accumulates into the pores of the stone, occluding them in a large extent. No (or very low) influence has been also noticed, even in very compact stone materials [73], measuring values comparable using either the hybrids or the polymer alone [79]. In other cases, unexpected increase in water vapor permeability have been found after the treatment [85]. This latter result is common in highly hydrophilic surfaces [86] and has been explained as related to a lower condensation phenomenon on the hydrophobic pore walls. Consequently, the diffusion rate of water vapor through hydrophobic pores is higher than the diffusion rate through hydrophilic pores. This result is neutralized when the applied hydrophobic product tends to fill the pores rather than coating them as a thin layer [73,87]. The consequent reduction in pores' dimensions slows down (or inhibits) the vapor flow [88]. This latter effect is dominant and a permeability decrease is observed as overall result [73].

Stone surfaces treated with the nano-TiO_2 polymer composites show higher photocatalytic activity under light irradiation in terms of organic dye discoloration (rhodamine B [73,74,79], methylene blue [76,81], methyl orange [80], methyl red [85]) in comparison to the untreated ones or to the polymer coating alone [74,79]. In those cases, it is evident that the TiO_2 nano-particles accelerate the oxidative degradation of the colorant. The amount of TiO_2 nano-filler on the surface is the main factor affecting the photodegradation efficacy [76]. Also where a defective surface coverage of the substrate, with aggregation of the nano-particles, has been observed, poor performance in photoactivity and inferior colorant discoloration are obtained [74]. This capability of self-cleaning from stain has been applied for the discoloration of graffiti drawn on modified coatings [83].

Photocathalytic nano-TiO_2 under light irradiation also exhibit super-hydrophilicity, which can interfere with the hydrophobicity induced by the polymer. In addition, the polymer degradation can take place, thus leading to a significant alteration of the original features.

Surfaces coated with acrylic/nano-TiO_2 hybrid usually reach lower performance and exhibit higher wettability because of the photo-degradation the acrylic component under UV radiation [75,76], while improved performances and durability have been observed using fluorine-based polymers [75]. Actually, the UV exposure of nano-TiO_2 fluorinated hybrids can yield a decrease of the contact angles. However, after the removal of any light source, the original values are recovered over time [81]. These results have been explained as due to the presence of an unnoticed film of water induced by the super-hydrophilicity of TiO_2 nano-particles, which remained on the stone surface for an extended time after its exposure to UV radiation, rather than to the degradation of the polymeric film [79].

The application of hybrid coatings have produced a better protection against salts formation, in comparison to that obtained applying the polymer, after exposure tests in a typical urban polluted environment [78] in wintertime. In particular the efficiency of the hybrid application is evident in decreasing concentration of sulphate and chloride ions [78].

The option to prevent the bio-colonization of restored surfaces is of high importance and has gained attention due to the fact that products used for conservation purposes may serve themselves as nutrient and thus support the biological colonization of the restored surfaces [82]. In this field, TiO_2 has been extensively used as biocide against various microorganisms (bacteria, fungi and viruses) due to its broad-spectrum antibiosis, its chemical stability, non-toxicity, high photo-reactivity, and cheapness [77,82]. Treatments with nano-TiO_2 hybrids are able to induce inhibition in biological growth. Although bioreceptivity of stone materials is highly variable and strongly depends on surface roughness, initial porosity and mineralogical characteristics, the positive effect has been observed on dissimilar lithotypes [76], while greater amounts of nano-powder have not improved the efficiency as biocidal agent.

5.3. Nano-TiO$_2$ Hybrids on Wood

During outdoor exposure, the initially appealing color of the wood surface changes very rapidly after a quite short period due to photo-oxidation processes, increase in hydrophilicity, erosion, and microbiological attack. Despite the fact that the loss of mechanical properties of weathered wood is limited to the upper cell rows and therefore negligible for construction purposes, the optical appearance is strongly altered [89]. The polymer lignin undergo bond cleavage and hydrogen abstraction, resulting in the formation of radicals or peroxides easily decomposed to produce chromophoric groups (such as carbonyl and carboxyl groups), and directly responsible for the wood yellowing/darkening [90].

The addition of TiO$_2$ nano-particles to coatings for wood such as UV-blocking agents is an extensive practice. From the different polymorphs present in TiO$_2$ nano-particles, rutile offers greater effectiveness as a stabilizer of photo-oxidation of polymers [91,92]. Rutile is an excellent candidate for UV protection because of its high refractive index (n = 2.7), wide band gap (3.05 eV), ability to scatter solar radiation and absorb UV radiation.

In applications for wood protection, many positive effects have been observed: the cellular microstructures of wood are easily and well coated by the nano-particles [93]; good anchoring takes place through interaction of TiO$_2$ particles with hemicellulose and lignin [94]; pronounced hydrophobicity [93,94] and reduction of fungi growth [95,96] are found. However, TiO$_2$ for UV protection of wood difficult work effectively if used alone or if the applied coating does not achieve a suitable thickness [97,98]. To improve the performances, the addition of a dopant (e.g., Ce [89]) or an organic commercial UV adsorber [98], as well as the preparation of an organic-inorganic hybrid have been employed.

Hybrids based on acrylic and methacrylic monomers have been successfully applied for protection of wood, where contradictory requirements have to be satisfied. In such application, the coating must tolerate the expansion and contraction of wood, influenced by humidity and temperature, it must be hard enough to withstand mechanical or chemical attack on the surface and stress at the interface due to dimensional changes of wood [89,91].

The TiO$_2$ nano-particles in the hybrid show a tendency toward agglomeration, therefore the distribution of nano-particles was often not homogenous throughout the matrix. Adhesion and water absorption were not significantly affected by the addition of TiO$_2$. Indeed, the addition of TiO$_2$ favors the absorption of a great amount of UV rays, thus increasing durability. This behavior was evident in wood samples subjected to either artificial or natural weathering, where degradation and color changes were observed [7,91,99].

6. SiO$_2$-TiO$_2$ Based Hybrids

In a more recent approach, TiO$_2$ is mixed with SiO$_2$ to obtain nano-particles for hybrid coatings.

Hybrids based on SiO$_2$-TiO$_2$ nano-particles can be obtained by mixing Ti-alkoxide and Si-alkoxide precursors, by grafting pre-synthesized silica with Ti-alkoxide or by mixing pre-formed titania colloidal nano-particles in a Si-alkoxide sol [100]. The addition of the polymer is the subsequent step, in general by using ultrasonic bath [101].

Acrylic polymers have been used as the organic component [102]. The employ of polydimethylsiloxane (PDMS) has provided hydrophobic properties and enhancement of toughness and flexibility of the silica network, thus preventing the gel from cracking during drying [103,104].

Oxalic acid has been used in order to catalyze the hydrolysis of TEOS, as well as both to control the grain growth of TiO$_2$ and the solvent drying, and to create a pH environment that prevents the particle TiO$_2$ agglomeration [103]. In this case, a transparent and cracked-free film derived from the sol-gel hybrid synthesized material is observed by SEM examinations. From TEM investigations, particle size distribution presents lower values at lower pH levels [104]. The addition of oxalic acid seems play a valuable role. In fact, in absence of this substance, coatings exhibiting cracks, a non-homogeneous distribution and aggregation of the TiO_2 nano-particles have been obtained [101]. These features negatively affect the performance of the treated surface, in particular the water absorption by capillarity.

To prevent cracking of the coating and to enhance photocatalytic activity by creating a mesoporous nano-composite, TiO_2 colloidal particles are added in the presence of a surfactant (n-octylamine [105]) to a silica oligomer, which is capable of adhering well to the stone. The employment of a nonionic surfactant allows avoiding additional steps (e.g., calcinations), which in Cultural Heritage field must be avoided because the coating will normally have to be applied in situ, on the external surface of a building stone material.

It has been demonstrated that n-octylamine, creating a mesoporous structure, efficiently reduces the capillary pressure and crack-free, therefore homogeneous, and coarser coatings on the stone surface are obtained [105,106]. In addition, favoring the interaction between the siloxane and the apolar carbonate stone, n-octylamine has enhanced the consolidant effectiveness of siloxane, while it is known that silicon-based products are less effective in pure carbonate stones [106]. However, color changes have been observed using n-octylamine, which may imparts a slightly yellowish color to the treated stone materials. This disadvantage is progressively reduced as the TiO_2 content is raised, since these particles have a whitening effect [105].

In the SiO_2–TiO_2 hybrids, the TiO_2 nano-particles promote a significant increase in viscosity [100]. An unexpected trend, in which the increase in particle size seems to produce lower viscosity, has been also observed. This behavior has been attributed to the agglomeration of TiO_2 nano-particles. In fact, aggregation takes place in order to reduce the high surface energy of the particles, which is more significant for the lowest average particle size.

The polymerization of the hybrid material can be assessed by observing changes in several absorptions in the FTIR spectra [103]. The incorporation of TiO_2 to the silica network is revealed by the creation of Si–O–Ti bonds, showing signals located in the spectral range of 920–950 cm^{-1}. The absence of the absorption of the Ti–O bonds of pure TiO_2 around 550 cm^{-1} can further support the realization of a proper hybrid structure [103].

TiO_2 nano-particles integrated into a SiO_2-based matrix, which is able of adhering firmly to the stone, provide long-term wear resistance, improve surface mechanical resistance, and salt crystallization degradation, along with self-cleaning and hydrophobic properties of the treated stone material [99,104,105].

The organic dye discoloration (methylene blue [103], methyl orange [104]) and the biofilm removal on treated stone samples support the self-cleaning properties of the coatings.

The organic component reduces the surface energy, producing hydrophobicity, as it is proved by the water capillary coefficient and contact angle measurements [103]; higher contact angle values are due to a higher surface roughness induced by the nano-particles in the hybrid [104]. Finally, the variation of the water vapor permeability and color parameters ranged within acceptable limits [103,104].

An additional valuable advantage in using SiO_2–TiO_2 hybrids has been recognized in the decreased release of nano-particles in the environment. This undoubtedly represents a positive response in terms of reduction of potential hazard of this class of materials [107].

7. Other Nano-Particles for Hybrids

In addition to the nano-composites described in the previous sections, other hybrids based on nano-particles have been proposed for applications on materials of the Cultural Heritage. The most common products are: nano-calcite in acrylic matrices [108,109], chemically more compatible with carbonate stones, as an alternative to silica-based treatments; nano-silver particles [110] or Cu-nano-particles [111] in acrylate/methacrylate polymers, to impart antimicrobial properties; silica-calcium oxalate hybrids as consolidants [112]; nano-hydroxyapatite in silane/siloxane matrices for stone consolidation and protection [87,113]; cellulose nano-crystals in UV-light curable in siloxane-modified methacrylic resin for wood protection [6].

Although less widely employed, the investigated products exhibit good compatibility with the stone/wood substrates, improved surface properties in terms of hydrophobicity/water absorption and strengthening of the stone material, as well as resistance to weathering.

8. Conclusions

Hybrid nano-composites, being carefully selected the nano-particles, the polymer matrix and the processes of preparation, are able to impart interesting properties and suitable performances to the treated stone/wood substrates.

These novel nano-materials have been deeply investigated and applied for conservative purposes in the field of the Cultural Heritage preservation. However, there is currently little known about their efficacy, behavior, and durability under actual conditions. In this regard, thorough studies and experimental activity should be performed directly on real surfaces. In addition, risks for human health and environment from the use of the new nano-materials for materials' preservation should be a primary concern in future research. Finally, in the literature thus reviewed, little or very low attention has been paid to economic considerations: the cost/performance ratio should be always taken into account and put in relation with the value of the artwork/structure to be preserved.

Funding: This research received no external funding.

Conflicts of Interest: The authors declare no conflict of interest.

References

1. Doehne, E.; Price, C.A. *Stone Conservation: An Overview of Current Research*, 2nd ed.; Getty Conservation Institute: Los Angeles, CA, USA, 2010.
2. Sierra-Fernandez, A.; Gomez-Villalba, L.S.; Rabanal, M.E.; Fort, R. New nanomaterials for applications in conservation and restoration of stony materials: A review. *Mater. Constr.* **2017**, *67*, 107. [CrossRef]
3. Cao, Y.; Salvini, A.; Camaiti, M. Oligoamide grafted with perfluoropolyether blocks: A potential protective coating for stone materials. *Prog. Org. Coat.* **2017**, *111*, 164–174. [CrossRef]
4. Corcione, C.E.; Manno, R.; Frigione, M. Sunlight curable boehmite/siloxane-modified methacrylic nano-composites: An innovative solution for the protection of carbonate stones. *Prog. Org. Coat.* **2016**, *97*, 222–232. [CrossRef]
5. Horie, C.V. *Materials for Conservation: Organic Consolidants, Adhesives and Coatings*; Routledge: London, UK, 2010.
6. Cataldi, A.; Corcione, C.E.; Frigione, M.; Pegoretti, A. Photocurable resin/nanocellulose composite coatings for wood protection. *Prog. Org. Coat.* **2017**, *106*, 128–136. [CrossRef]
7. Fufa, S.M.; Jelle, B.P.; Hovde, P.J.; Rørvik, P.M. Coated wooden claddings and the influence of nanoparticles on the weathering performance. *Prog. Org. Coat.* **2012**, *75*, 72–78. [CrossRef]
8. Turri, S.; Alborghetti, L.; Levi, M. Formulation and properties of a model two-component nanocomposite coating from organophilic nanoclays. *J. Polym. Res.* **2008**, *15*, 365–372. [CrossRef]
9. Landry, V.; Blanchet, P.; Riedl, B. Mechanical and optical properties of clay-based nanocomposites coatings for wood flooring. *Prog. Org. Coat.* **2010**, *67*, 381–388. [CrossRef]
10. Yu, Y.-H.; Lin, C.-Y.; Yeh, J.-M.; Lin, W.-H. Preparation and properties of poly(vinyl alcohol)-clay nanocomposite materials. *Polymer* **2003**, *44*, 3553–3560. [CrossRef]
11. Corcione, C.E.; Frigione, M. Characterization of nanocomposites by thermal analysis. *Materials* **2012**, *5*, 2960–2980. [CrossRef]
12. Decker, C.; Keller, L.; Zahouily, K.; Benfarhi, S. Synthesis of nanocomposite polymers by UV-radiation curing. *Polym. Blends Compos. Hybrid Polym. Mater.* **2005**, *46*, 6640–6648. [CrossRef]
13. Bellisario, D.; Quadrini, F.; Santo, L. Nano-clay filled polyester coatings. *Prog. Org. Coat.* **2013**, *76*, 1863–1868. [CrossRef]
14. Corcione, C.E.; Frigione, M. UV-cured polymer-boehmite nanocomposite as protective coating for wood elements. *Prog. Org. Coat.* **2012**, *74*, 781–787. [CrossRef]

15. Ray, S.S.; Okamoto, M. Polymer/layered silicate nanocomposites: A review from preparation to processing. *Prog. Polym. Sci.* **2003**, *28*, 1539–1641. [CrossRef]
16. Bhattacharyya, K.G.; Gupta, S.S. Adsorption of a few heavy metals on natural and modified kaolinite and montmorillonite: A review. *Adv. Colloid Interface Sci.* **2008**, *140*, 114–131. [CrossRef] [PubMed]
17. Alexandre, M.; Dubois, P. Polymer-layered silicate nanocomposites: Preparation, properties and uses of a new class of materials. *Mater. Sci. Eng. R Rep.* **2000**, *28*, 1–63. [CrossRef]
18. Wang, S.; Zhang, Y.; Ren, W.; Zhang, Y.; Lin, H. Morphology, mechanical and optical properties of transparent BR/clay nanocomposites. *Polym. Test.* **2005**, *24*, 766–774. [CrossRef]
19. Nkeuwa, W.N.; Riedl, B.; Landry, V. UV-cured clay/based nanocomposite topcoats for wood furniture. Part II: Dynamic viscoelastic behavior and effect of relative humidity on the mechanical properties. *Prog. Org. Coat.* **2014**, *77*, 12–23. [CrossRef]
20. D'Arienzo, L.; Scarfato, P.; Incarnato, L. New polymeric nanocomposites for improving the protective and consolidating efficiency of tuff stone. *J. Cult. Herit.* **2008**, *9*, 253–260. [CrossRef]
21. Licchelli, M.; Malagodi, M.; Weththimuni, M.; Zanchi, C. Anti-graffiti nanocomposite materials for surface protection of a very porous stone. *Appl. Phys. A* **2014**, *116*, 1525–1539. [CrossRef]
22. Ocak, Y.; Sofuoglu, A.; Tihminlioglu, F.; Böke, H. Sustainable bio-nano composite coatings for the protection of marble surfaces. *J. Cult. Herit.* **2015**, *16*, 299–306. [CrossRef]
23. Nkeuwa, W.N.; Riedl, B.; Landry, V. UV-cured clay/based nanocomposite topcoats for wood furniture: Part I: Morphological study, water vapor transmission rate and optical clarity. *Prog. Org. Coat.* **2014**, *77*, 1–11. [CrossRef]
24. Fufa, S.M.; Jelle, B.P.; Hovde, P.J. Effects of TiO$_2$ and clay nanoparticles loading on weathering performance of coated wood. *Prog. Org. Coat.* **2013**, *76*, 1425–1429. [CrossRef]
25. Fufa, S.M.; Jelle, B.P.; Hovde, P.J. Weathering performance of spruce coated with water based acrylic paint modified with TiO$_2$ and clay nanoparticles. *Prog. Org. Coat.* **2013**, *76*, 1543–1548. [CrossRef]
26. Karger-Kocsis, J.; Lendvai, L. Polymer/boehmite nanocomposites: A review. *J. Appl. Polym. Sci.* **2017**, *135*, 45573. [CrossRef]
27. Corcione, C.E.; Frigione, M. Cure kinetics and physical characterization of epoxy/modified boehmite nanocomposites. *J. Adhes. Sci. Technol.* **2017**, *31*, 645–662. [CrossRef]
28. Corcione, C.E.; Frigione, M.; Maffezzoli, A.; Malucelli, G. Photo–DSC and real time–FT-IR kinetic study of a UV curable epoxy resin containing o-Boehmites. *Eur. Polym. J.* **2008**, *44*, 2010–2023. [CrossRef]
29. Corcione, C.E.; Frigione, M.; Acierno, D. Rheological characterization of UV-curable epoxy systems: Effects of o-Boehmite nanofillers and a hyperbranched polymeric modifier. *J. Appl. Polym. Sci.* **2009**, *112*, 1302–1310. [CrossRef]
30. Corcione, C.E.; Frigione, M. Surface characterization of novel hydrophobic UV-curable siloxane-modified methacrylate/boehmite nanocomposites. *Polym. Compos.* **2013**, *34*, 1546–1552. [CrossRef]
31. Corcione, C.E.; Manno, R.; Frigione, M. Sunlight-curable boehmite/siloxane-modified methacrylic based nanocomposites as insulating coatings for stone substrates. *Prog. Org. Coat.* **2016**, *95*, 107–119. [CrossRef]
32. Corcione, C.E.; De Simone, N.; Santarelli, M.L.; Frigione, M. Protective properties and durability characteristics of experimental and commercial organic coatings for the preservation of porous stone. *Prog. Org. Coat.* **2017**, *103*, 193–203. [CrossRef]
33. Corcione, C.E.; Frigione, M. Novel UV-cured nanocomposite used for the protection of walnut wood artworks. *Wood Res.* **2014**, *59*, 229–244.
34. Cappelletti, G.; Fermo, P. Hydrophobic and superhydrophobic coatings for limestone and marble conservation. In *Smart Composite Coatings and Membranes*; Montemor, M.F., Ed.; Elsevier Inc.: Amsterdam, The Netherlands, 2016; pp. 421–452.
35. Chattopadhyay, D.K.; Raju, K.V.S.N. Structural engineering of polyurethane coatings for high performance applications. *Prog. Polym. Sci. Oxf.* **2007**, *32*, 352–418. [CrossRef]
36. Ruffolo, S.A.; La Russa, M.F.; Ricca, M.; Belfiore, C.M.; Macchia, A.; Comite, V.; Pezzino, A.; Crisci, G.M. New insights on the consolidation of salt weathered limestone: The case study of Modica stone. *Bull. Eng. Geol. Environ.* **2017**, *76*, 11–20. [CrossRef]
37. Aslanidou, D.; Karapanagiotis, I.; Panayiotou, C. Tuning the wetting properties of siloxane-nanoparticle coatings to induce superhydrophobicity and superoleophobicity for stone protection. *Mater. Des.* **2016**, *108*, 736–744. [CrossRef]

38. De Ferri, L.; Lottici, P.P.; Lorenzi, A.; Montenero, A.; Salvioli-Mariani, E. Study of silica nanoparticles—Polysiloxane hydrophobic treatments for stone-based monument protection. *J. Cult. Herit.* **2011**, *12*, 356–363. [CrossRef]
39. Chatzigrigoriou, A.; Manoudis, P.N.; Karapanagiotis, I. Fabrication of water repellent coatings using waterborne resins for the protection of the cultural heritage. *Macromol. Symp.* **2013**, *331–332*, 158–165. [CrossRef]
40. Aslanidou, D.; Karapanagiotis, I.; Lampakis, D. Waterborne superhydrophobic and superoleophobic coatings for the protection of marble and sandstone. *Materials* **2018**, *11*, 585. [CrossRef] [PubMed]
41. Facio, D.S.; Ordoñez, J.A.; Almoraima Gil, M.L.; Carrascosa, L.A.M.; Mosquera, M.J. New consolidant-hydrophobic treatment by combining SiO_2 composite and fluorinated alkoxysilane: Application on decayed biocalcareous stone from an 18th century cathedral. *Coatings* **2018**, *8*, 170. [CrossRef]
42. Jacobs, D.S.; Huang, S.-R.; Cheng, Y.-L.; Rabb, S.A.; Gorham, J.M.; Krommenhoek, P.J.; Yu, L.L.; Nguyen, T.; Sung, L. Surface degradation and nanoparticle release of a commercial nanosilica/polyurethane coating under UV exposure. *J. Coat. Technol. Res.* **2016**, *13*, 735–751. [CrossRef] [PubMed]
43. Pagliolico, S.L.; Ozzello, E.D.; Sassi, G.; Bongiovanni, R. Characterization of a hybrid nano-silica waterborne polyurethane coating for clay bricks. *J. Coat. Technol. Res.* **2016**, *13*, 267–276. [CrossRef]
44. Sow, C.; Riedl, B.; Blanchet, P. Kinetic studies of UV-waterborne nanocomposite formulations with nanoalumina and nanosilica. *Prog. Org. Coat.* **2010**, *67*, 188–194. [CrossRef]
45. Corcione, C.E.; Striani, R.; Frigione, M. Organic-inorganic UV-cured methacrylic-based hybrids as protective coatings for different substrates. *Prog. Org. Coat.* **2014**, *77*, 1117–1125. [CrossRef]
46. Corcione, C.E.; Striani, R.; Capone, C.; Molfetta, M.; Vendetta, S.; Frigione, M. Preliminary study of the application of a novel hydrophobic photo-polymerizable nano-structured coating on concrete substrates. *Prog. Org. Coat.* **2018**, *121*, 182–189. [CrossRef]
47. Morote-Martínez, V.; Pascual-Sánchez, V.; Martín-Martínez, J.M. Improvement in mechanical and structural integrity of natural stone by applying unsaturated polyester resin-nanosilica hybrid thin coating. *Eur. Polym. J.* **2008**, *44*, 3146–3155. [CrossRef]
48. Nguyen, T.; Pellegrin, B.; Bernard, C.; Rabb, S.; Stuztman, P.; Gorham, J.M.; Gu, X.; Yu, L.L.; Chin, J.W. Characterization of surface accumulation and release of nanosilica during irradiation of polymer nanocomposites by ultraviolet light. *J. Nanosci. Nanotechnol.* **2012**, *12*, 6202–6215. [CrossRef] [PubMed]
49. Sung, L.; Stanley, D.; Gorham, J.M.; Rabb, S.; Gu, X.; Yu, L.L.; Nguyen, T. A quantitative study of nanoparticle release from nanocoatings exposed to UV radiation. *J. Coat. Technol. Res.* **2014**, *12*, 121–135. [CrossRef]
50. Stefanidou, M.; Matziaris, K.; Karagiannis, G. Hydrophobization by means of nanotechnology on greek sandstones used as building facades. *Geosci. Switz.* **2013**, *3*, 30–45. [CrossRef]
51. Fic, S.; Szewczak, A.; Barnat-Hunek, D.; Lagód, G. Processes of fatigue destruction in nanopolymer-hydrophobised ceramic bricks. *Materials* **2017**, *10*, 44. [CrossRef] [PubMed]
52. Manoudis, P.N.; Tsakalof, A.; Karapanagiotis, I.; Zuburtikudis, I.; Panayiotou, C. Fabrication of super-hydrophobic surfaces for enhanced stone protection. *Surf. Coat. Technol.* **2009**, *203*, 1322–1328. [CrossRef]
53. Mosquera, M.J.; De Los Santos, D.M.; Rivas, T.; Sanmartín, P.; Silva, B. New nanomaterials for protecting and consolidating stone. *J. Nano Res.* **2009**, *8*, 1–12. [CrossRef]
54. Mosquera, M.J.; De Los Santos, D.M.; Rivas, T. Surfactant-synthesized ormosils with application to stone restoration. *Langmuir* **2010**, *26*, 6737–6745. [CrossRef] [PubMed]
55. Li, D.; Xu, F.; Liu, Z.; Zhu, J.; Zhang, Q.; Shao, L. The effect of adding PDMS-OH and silica nanoparticles on sol–gel properties and effectiveness in stone protection. *Appl. Surf. Sci.* **2013**, *266*, 368–374. [CrossRef]
56. Simionescu, B.; Olaru, M.; Aflori, M.; Cotofana, C. Silsesquioxane-based hybrid nanocomposite with self-assembling properties for porous limestones conservation. *High Perform. Polym.* **2010**, *22*, 42–55. [CrossRef]
57. Zornoza-Indart, A.; Lopez-Arce, P.; Leal, N.; Simão, J.; Zoghlami, K. Consolidation of a Tunisian bioclastic calcarenite: From conventional ethyl silicate products to nanostructured and nanoparticle based consolidants. *Constr. Build. Mater.* **2016**, *116*, 188–202. [CrossRef]
58. Bailly, M.; Kontopoulou, M.; El Mabrouk, K. Effect of polymer/filler interactions on the structure and rheological properties of ethylene-octene copolymer/nanosilica composites. *Polymer* **2010**, *51*, 5506–5515. [CrossRef]

59. Hao, X.; Kaschta, J.; Pan, Y.; Liu, X.; Schubert, D.W. Intermolecular cooperativity and entanglement network in a miscible PLA/PMMA blend in the presence of nanosilica. *Polymer* **2016**, *82*, 57–65. [CrossRef]
60. Lepcio, P.; Ondreas, F.; Zarybnicka, K.; Zboncak, M.; Caha, O.; Jancar, J. Bulk polymer nanocomposites with preparation protocol governed nanostructure: The origin and properties of aggregates and polymer bound clusters. *Soft Matter* **2018**, *14*, 2094–2103. [CrossRef] [PubMed]
61. Esposito Corcione, C.; Striani, R.; Frigione, M. Hydrophobic photopolymerizable nanostructured hybrid materials: An effective solution for the protection of porous stones. *Polym. Compos.* **2015**, *36*, 1039–1047. [CrossRef]
62. Vecchiattini, R.; Fratini, F.; Rescic, S.; Riminesi, C.; Mauri, M.; Vicini, S. The marly limestone, a difficult material to restore: The case of the San Fruttuoso di Capodimonte Abbey (Genoa, Italy). *J. Cult. Herit.* **2018**, in press. [CrossRef]
63. Facio, D.S.; Mosquera, M.J. Simple strategy for producing superhydrophobic nanocomposite coatings in situ on a building substrate. *ACS Appl. Mater. Interfaces* **2013**, *5*, 7517–7526. [CrossRef] [PubMed]
64. Corcione, C.E.; Striani, R.; Frigione, M. Novel hydrophobic free-solvent UV-cured hybrid organic–inorganic methacrylic-based coatings for porous stones. *Prog. Org. Coat.* **2014**, *77*, 803–812. [CrossRef]
65. Sow, C.; Riedl, B.; Blanchet, P. UV-waterborne polyurethane-acrylate nanocomposite coatings containing alumina and silica nanoparticles for wood: Mechanical, optical, and thermal properties assessment. *J. Coat. Technol. Res.* **2011**, *8*, 211–221. [CrossRef]
66. Nkeuwa, W.N.; Riedl, B.; Landry, V. Wood surfaces protected with transparent multilayer UV-cured coatings reinforced with nanosilica and nanoclay. Part I: Morphological study and effect of relative humidity on adhesion strength. *J. Coat. Technol. Res.* **2014**, *11*, 283–301. [CrossRef]
67. Nkeuwa, W.N.; Riedl, B.; Landry, V. Wood surfaces protected with transparent multilayer UV-cured coatings reinforced with nanosilica and nanoclay. Part II: Application of a standardized test method to study the effect of relative humidity on scratch resistance. *J. Coat. Technol. Res.* **2014**, *11*, 993–1011. [CrossRef]
68. Nikolic, M.; Lawther, J.M.; Sanadi, A.R. Use of nanofillers in wood coatings: A scientific review. *J. Coat. Technol. Res.* **2015**, *12*, 445–461. [CrossRef]
69. Kumar, A.; Petrič, M.; Kričej, B.; Žigon, J.; Tywoniak, J.; Hajek, P.; Škapin, A.S.; Pavlič, M. Liquefied-wood-based polyurethane-nanosilica hybrid coatings and hydrophobization by self-assembled monolayers of orthotrichlorosilane (OTS). *ACS Sustain. Chem. Eng.* **2015**, *3*, 2533–2541. [CrossRef]
70. Chu, Z.; Seeger, S. Robust superhydrophobic wood obtained by spraying silicone nanoparticles. *RSC Adv.* **2015**, *5*, 21999–22004. [CrossRef]
71. Munafò, P.; Goffredo, G.B.; Quagliarini, E. TiO$_2$-based nanocoatings for preserving architectural stone surfaces: An overview. *Constr. Build. Mater.* **2015**, *84*, 201–218. [CrossRef]
72. Quagliarini, E.; Graziani, L.; Diso, D.; Licciulli, A.; D'Orazio, M. Is nano-TiO$_2$ alone an effective strategy for the maintenance of stones in Cultural Heritage? *J. Cult. Herit.* **2018**, *30*, 81–91. [CrossRef]
73. Gherardi, F.; Roveri, M.; Goidanich, S.; Toniolo, L. Photocatalytic nanocomposites for the protection of European architectural heritage. *Materials* **2018**, *11*, 65. [CrossRef] [PubMed]
74. Gherardi, F.; Goidanich, S.; Toniolo, L. Improvements in marble protection by means of innovative photocatalytic nanocomposites. *Prog. Org. Coat.* **2018**, *121*, 13–22. [CrossRef]
75. La Russa, M.F.; Rovella, N.; Alvarez de Buergo, M.; Belfiore, C.M.; Pezzino, A.; Crisci, G.M.; Ruffolo, S.A. Nano-TiO$_2$ coatings for cultural heritage protection: The role of the binder on hydrophobic and self-cleaning efficacy. *Prog. Org. Coat.* **2016**, *91*, 1–8. [CrossRef]
76. La Russa, M.F.; Ruffolo, S.A.; Rovella, N.; Belfiore, C.M.; Palermo, A.M.; Guzzi, M.T.; Crisci, G.M. Multifunctional TiO$_2$ coatings for cultural heritage. *Prog. Org. Coat.* **2012**, *74*, 186–191. [CrossRef]
77. Aflori, M.; Simionescu, B.; Bordianu, I.-E.; Sacarescu, L.; Varganici, C.-D.; Doroftei, F.; Nicolescu, A.; Olaru, M. Silsesquioxane-based hybrid nanocomposites with methacrylate units containing titania and/or silver nanoparticles as antibacterial/antifungal coatings for monumental stones. *Mater. Sci. Eng. B* **2013**, *178*, 1339–1346. [CrossRef]
78. Cappelletti, G.; Fermo, P.; Camiloni, M. Smart hybrid coatings for natural stones conservation. *Prog. Org. Coat.* **2015**, *78*, 511–516. [CrossRef]
79. Colangiuli, D.; Calia, A.; Bianco, N. Novel multifunctional coatings with photocatalytic and hydrophobic properties for the preservation of the stone building heritage. *Constr. Build. Mater.* **2015**, *93*, 189–196. [CrossRef]

80. D'Orazio, L.; Grippo, A. A water dispersed Titanium dioxide/poly(carbonate urethane) nanocomposite for protecting cultural heritage: Preparation and properties. *Prog. Org. Coat.* **2015**, *79*, 1–7. [CrossRef]
81. Alfieri, I.; Lorenzi, A.; Ranzenigo, L.; Lazzarini, L.; Predieri, G.; Lottici, P.P. Synthesis and characterization of photocatalytic hydrophobic hybrid TiO_2–SiO_2 coatings for building applications. *Build. Environ.* **2017**, *111*, 72–79. [CrossRef]
82. Barberio, M.; Veltri, S.; Sokullu, E.; Xu, F.; Gauthier, M.A.; Antici, P. Preparation and characterization of nanostructured films: study of hydrophobicity and antibacterial properties for surface protection. In *Advanced Processing and Manufacturing Technologies for Nanostructured and Multifunctional Materials II: A Collection of Papers Presented at the 39th International Conference on Advanced Ceramics and Composites*; Ohji, T., Singh, M., Halbig, M., Eds.; Wiley Blackwell: Hoboken, NJ, USA, 2015; pp. 101–111.
83. Scalarone, D.; Lazzari, M.; Chiantore, O. Acrylic protective coatings modified with titanium dioxide nanoparticles: Comparative study of stability under irradiation. *Polym. Degrad. Stab.* **2012**, *97*, 2136–2142. [CrossRef]
84. Milanesi, F.; Cappelletti, G.; Annunziata, R.; Bianchi, C.L.; Meroni, D.; Ardizzone, S. Siloxane–TiO_2 hybrid nanocomposites. The structure of the hydrophobic layer. *J. Phys. Chem. C* **2010**, *114*, 8287–8293. [CrossRef]
85. Petronella, F.; Pagliarulo, A.; Striccoli, M.; Calia, A.; Lettieri, M.; Colangiuli, D.; Curri, M.L.; Comparelli, R. Colloidal nanocrystalline semiconductor materials as photocatalysts for environmental protection of architectural stone. *Crystals* **2017**, *7*, 30. [CrossRef]
86. Kronlund, D.; Bergbreiter, A.; Meierjohann, A.; Kronberg, L.; Lindén, M.; Grosso, D.; Smått, J.-H. Hydrophobization of marble pore surfaces using a total immersion treatment method—Product selection and optimization of concentration and treatment time. *Prog. Org. Coat.* **2015**, *85*, 159–167. [CrossRef]
87. Luo, Y.; Xiao, L.; Zhang, X. Characterization of TEOS/PDMS/HA nanocomposites for application as consolidant/hydrophobic products on sandstones. *J. Cult. Herit.* **2015**, *16*, 470–478. [CrossRef]
88. Manoudis, P.N.; Karapanagiotis, I.; Tsakalof, A.; Zuburtikudis, I.; Kolinkeová, B.; Panayiotou, C. Superhydrophobic films for the protection of outdoor cultural heritage assets. *Appl. Phys. A* **2009**, *97*, 351–360. [CrossRef]
89. Forsthuber, B.; Müller, U.; Teischinger, A.; Grüll, G. Chemical and mechanical changes during photooxidation of an acrylic clear wood coat and its prevention using UV absorber and micronized TiO_2. *Polym. Degrad. Stab.* **2013**, *98*, 1329–1338. [CrossRef]
90. Guo, H.; Klose, D.; Hou, Y.; Jeschke, G.; Burgert, I. Highly efficient UV protection of the biomaterial wood by a transparent TiO_2/Ce xerogel. *ACS Appl. Mater. Interfaces* **2017**, *9*, 39040–39047. [CrossRef] [PubMed]
91. Moya, R.; Rodríguez-Zúñiga, A.; Vega-Baudrit, J.; Puente-Urbina, A. Effects of adding TiO_2 nanoparticles to a water-based varnish for wood applied to nine tropical woods of Costa Rica exposed to natural and accelerated weathering. *J. Coat. Technol. Res.* **2017**, *14*, 141–152. [CrossRef]
92. Sun, Q.; Lu, Y.; Zhang, H.; Zhao, H.; Yu, H.; Xu, J.; Fu, Y.; Yang, D.; Liu, Y. Hydrothermal fabrication of rutile TiO_2 submicrospheres on wood surface: An efficient method to prepare UV-protective wood. *Mater. Chem. Phys.* **2012**, *133*, 253–258. [CrossRef]
93. Rassam, G.; Abdi, Y.; Abdi, A. Deposition of TiO_2 nano-particles on wood surfaces for UV and moisture protection. *J. Exp. Nanosci.* **2012**, *7*, 468–476. [CrossRef]
94. Pori, P.; Vilčnik, A.; Petrič, M.; Škapin, A.S.; Mihelčič, M.; Šurca Vuk, A.; Novak, U.; Orel, B. Structural studies of TiO_2/wood coatings prepared by hydrothermal deposition of rutile particles from $TiCl_4$ aqueous solutions on spruce (Picea Abies) wood. *Appl. Surf. Sci.* **2016**, *372*, 125–138. [CrossRef]
95. Guo, H.; Bachtiar, E.V.; Ribera, J.; Heeb, M.; Schwarze, F.W.M.R.; Burgert, I. Non-biocidal preservation of wood against brown-rot fungi with a TiO_2/Ce xerogel. *Green Chem.* **2018**, *20*, 1375–1382. [CrossRef]
96. Oliva, R.; Salvini, A.; Di Giulio, G.; Capozzoli, L.; Fioravanti, M.; Giordano, C.; Perito, B. TiO_2-Oligoaldaramide nanocomposites as efficient core-shell systems for wood preservation. *J. Appl. Polym. Sci.* **2015**, *132*, 42047. [CrossRef]
97. Veronovski, N.; Verhovšek, D.; Godnjavec, J. The influence of surface-treated nano-TiO_2 (rutile) incorporation in water-based acrylic coatings on wood protection. *Wood Sci. Technol.* **2013**, *47*, 317–328. [CrossRef]
98. Saha, S.; Kocaefe, D.; Sarkar, D.K.; Boluk, Y.; Pichette, A. Effect of TiO_2-containing nano-coatings on the color protection of heat-treated jack pine. *J. Coat. Technol. Res.* **2011**, *8*, 183–190. [CrossRef]
99. Vlad Cristea, M.; Riedl, B.; Blanchet, P. Enhancing the performance of exterior waterborne coatings for wood by inorganic nanosized UV absorbers. *Prog. Org. Coat.* **2010**, *69*, 432–441. [CrossRef]

100. Pinho, L.; Mosquera, M.J. Photocatalytic activity of TiO$_2$–SiO$_2$ nanocomposites applied to buildings: Influence of particle size and loading. *Appl. Catal. B Environ.* **2013**, *134–135*, 205–221. [CrossRef]
101. Crupi, V.; Fazio, B.; Gessini, A.; Kis, Z.; La Russa, M.F.; Majolino, D.; Masciovecchio, C.; Ricca, M.; Rossi, B.; Ruffolo, S.A.; et al. TiO$_2$–SiO$_2$–PDMS nanocomposite coating with self-cleaning effect for stone material: Finding the optimal amount of TiO$_2$. *Constr. Build. Mater.* **2018**, *166*, 464–471. [CrossRef]
102. D'Amato, R.; Caneve, L.; Giancristofaro, C.; Persia, F.; Pilloni, L.; Rinaldi, A. Development of nanocomposites for conservation of artistic stones. *Proc. Inst. Mech. Eng. Part N J. Nanoeng. Nanosyst.* **2013**, *228*, 19–26. [CrossRef]
103. Kapridaki, C.; Maravelaki-Kalaitzaki, P. TiO$_2$–SiO$_2$–PDMS nano-composite hydrophobic coating with self-cleaning properties for marble protection. *Prog. Org. Coat.* **2013**, *76*, 400–410. [CrossRef]
104. Kapridaki, C.; Pinho, L.; Mosquera, M.J.; Maravelaki-Kalaitzaki, P. Producing photoactive, transparent and hydrophobic SiO$_2$-crystalline TiO$_2$ nanocomposites at ambient conditions with application as self-cleaning coatings. *Appl. Catal. B Environ.* **2014**, *156–157*, 416–427. [CrossRef]
105. Pinho, L.; Mosquera, M.J. Titania-silica nanocomposite photocatalysts with application in stone self-cleaning. *J. Phys. Chem. C* **2011**, *115*, 22851–22862. [CrossRef]
106. Pinho, L.; Elhaddad, F.; Facio, D.S.; Mosquera, M.J. A novel TiO$_2$–SiO$_2$ nanocomposite converts a very friable stone into a self-cleaning building material. *Appl. Surf. Sci.* **2013**, *275*, 389–396. [CrossRef]
107. Ortelli, S.; Poland, C.A.; Baldi, G.; Costa, A.L. Silica matrix encapsulation as a strategy to control ROS production while preserving photoreactivity in nano-TiO$_2$. *Environ. Sci. Nano* **2016**, *3*, 602–610. [CrossRef]
108. Coltelli, M.-B.; Paolucci, D.; Castelvetro, V.; Bianchi, S.; Mascha, E.; Panariello, L.; Pesce, C.; Weber, J.; Lazzeri, A. Preparation of water suspensions of nanocalcite for Cultural Heritage applications. *Nanomaterials* **2018**, *8*, 254. [CrossRef] [PubMed]
109. Aldoasri, A.M.; Darwish, S.S.; Adam, A.M.; Elmarzugi, A.N.; Ahmed, M.S. Enhancing the durability of calcareous stone monuments of ancient Egypt using CaCO$_3$ nanoparticles. *Sustainability* **2017**, *9*, 1392. [CrossRef]
110. Essa, A.M.M.; Khallaf, M.K. Biological nanosilver particles for the protection of archaeological stones against microbial colonization. *Int. Biodeterior. Biodegrad.* **2014**, *94*, 31–37. [CrossRef]
111. Essa, A.M.M.; Khallaf, M.K. Antimicrobial potential of consolidation polymers loaded with biological copper nanoparticles. *BMC Microbiol.* **2016**, *16*, 144. [CrossRef] [PubMed]
112. Verganelaki, A.; Kilikoglou, V.; Karatasios, I.; Maravelaki-Kalaitzaki, P. A biomimetic approach to strengthen and protect construction materials with a novel calcium-oxalate–silica nanocomposite. *Constr. Build. Mater.* **2014**, *62*, 8–17. [CrossRef]
113. Maravelaki, P.; Verganelaki, A. A hybrid consolidant of nano-hydroxyapatite and silica inspired from patinas for stone conservation. In *Advanced Materials for the Conservation of Stone*; Hosseini, M., Karapanagiotis, I., Eds.; Springer International Publishing: Cham, Switzerland, 2018; pp. 79–95.

© 2018 by the authors. Licensee MDPI, Basel, Switzerland. This article is an open access article distributed under the terms and conditions of the Creative Commons Attribution (CC BY) license (http://creativecommons.org/licenses/by/4.0/).

MDPI
St. Alban-Anlage 66
4052 Basel
Switzerland
Tel. +41 61 683 77 34
Fax +41 61 302 89 18
www.mdpi.com

Coatings Editorial Office
E-mail: coatings@mdpi.com
www.mdpi.com/journal/coatings